On the Shoulders of Giants

10 Quantum Pioneers of the Past

Brian T. Lenahan &
Kenna Hughes-Castleberry

Acclaim for Brian Lenahan

"Brian Lenahan is truly a 'force of nature' in the Quantum World'" - *Deborah Berebichez, Ph.D., Physicist, TV Host, and Leader in STEM*

"I encourage people to buy Quantum Boost and Quantum Excellence to understand the technology and what to do with it. It's a clever explanation and easily read… Brian has a gift for explaining complex topics with simple words." - *Christophe Pere, PhD.*

"Clearly, you are one of the top voices in Quantum. We need people to continue demystifying the technology, play down the hype… and articulate real-world business cases" – *Louis-Pierre Gravelle, International IP Expert*

You are simply THE BEST! Congratulations Brian! You are an inspiration! - *Quantum Empire Founder & Director of Iconic Innovation*

"Brian Lenahan You are a leader in the space and few appreciate what life at 300 Qubits mean for the world and its future when we get there" – *Jake Tiley, Capital Markets Investor*

"Your journey as a leader in the quantum space has been very inspiring" – *Steven Wilson, Centre for Digital Transformation*

"Very forward-looking, Brian, and very needed by the industry. Excellent series." – *Mike Kilroy, EVP, HKA Communications on "Quantum as a Professional" series*

Brian Lenahan thank you for participating in the "bqb Quantum Linkedin TOP VOICES" ranking and congratulations on your well-deserved position. You do an incredible job of spreading quantum technologies – *Barcelona Qubits*

Acclaim for Kenna Hughes-Castleberry

"Kenna is a thorough interviewer - she focuses on the personal story." —Andy Weir, bestselling author of *the Martian, Artemis,* and *Project Hail Mary*

"Kenna Hughes-Castleberry is an accomplished science writer who is passionate about giving a voice to unknown scientists, particularly women and minorities, and people whose stories have never been told. I love her blogcast *From the Biblio-Files*, where she interviewed several popular science and science fiction writers about their work, and her short course *Science Communication in the wake of COVID-19*, where she explored how to improve the coverage of science. I look forward to reading her books and following her work!" *—Marianna Limas, Science Writing News Roundup*

"By focusing her lens on the people behind the science, Kenna Hughes-Castleberry captures more than just the thrill of discovery. With a keen eye, she portrays the self-doubt and sacrifice that scientists—especially the unsung heroes—experience, just like the rest of us. By telling their stories, Hughes-Castleberry makes the often obscure worlds of physics and computing tangible and real." *—Bill Gourgey, adjunct lecturer, Johns Hopkins University; contributing writer, Popular Science*

"Kenna Hughes-Castleberry's eloquent writing shines a needed light on brilliant, but hidden individuals in science and history." *—Sheril Kirshenbaum, best-selling author and scientist at Michigan State University*

"Kenna Hughes-Castleberry always displays a keen eye for reporting on the scientific innovations of today that will likely play a critical role in shaping tomorrow's world. As a science writer, her work in covering those innovations also helps sow the seeds for that bright future we hope science may help us achieve." *—Micah Hanks, Editor-in-Chief and Co-Founder of The Debrief*

"Kenna is a well-known technical analyst and reporter. She is highly equipped to cover the historic and scientific aspects of physics and computing. Kenna is an in-depth researcher as well as a wonderful judge of the human aspects of any scientific topic." *—Alan Meckler, Ph.D. CEO of 3DR Holdings LLC and Inside Quantum Technology*

"Kenna's ability to grasp and distill the science is matched only by her gifts as a storyteller. She has achieved a rare breakthrough in fusion by artfully marrying her deep technical knowledge with a compelling, vividly told, and highly readable narrative style."
—Rina Gill, Head of Communications at Unbound Publishing

"At *Inside Quantum Technology* Kenna has been both a superb writer and an insightful interviewer. So I am looking forward immensely to reading this book." *—Lawrence Gasman, President and Founder of Inside Quantum Technology*

Acclaim for On the Shoulders of Giants

"A remarkable history of quantum technology as seen through the stories of its forgotten heroes." —Chris Ferrie, quantum physicist and bestselling author of the *Baby University* book series and *Quantum Bullsh*t*

"To read about quantum physics is to explore some of the most profound and fascinating questions posed by contemporary science, and we do ourselves a disservice when we reduce the history of the field to a few familiar names. Kenna Hughes-Castleberry [and Brian Lenahan] look beyond the usual narrative, introducing us to people who have been an essential part of that vast collaborative voyage of discovery and whose stories are as diverse, complex, and deeply interesting as the science itself." —Robert Charles Wilson, best-selling science fiction author and Hugo award winner.

"While you may not recognize their names, you'll be inspired by their stories and never forget their contributions to physics and computing." *—Sheril Kirshenbaum, best-selling author and scientist at Michigan State University*

Also by Brian Lenahan on Amazon

Artificial Intelligence: Foundations for Business Leaders and Consultants - *2019*

Digital Coach: Coaching in the Era of Artificial Intelligence - *2019*

Artificial Intelligence: Transitions – How to Successfully Prepare for Career Where AI is Everywhere" - *2019*

Deep Health: Using Artificial Intelligence to Live Longer and Healthier" - *2020*

Quantum Boost: Using Quantum Computing to Supercharge Your Business - *2021*
(available in English, Spanish, Portuguese and French)

Quantum Excellence: How Leading Organizations are Deploying the Transformational Technology – *2022*
(available in English & Spanish)

On the Shoulders of Giants

10 Quantum Pioneers of the Past

Brian T. Lenahan &
Kenna Hughes-Castleberry

Aquitaine Innovation Advisors
Toronto

www.aquitaineinnovationadvisors.com

Designed by Brian T. Lenahan

Library of Congress Cataloging-in-Publication Data
Names: Lenahan, Brian T., 1963- author; Hughes-Castleberry, Kenna, 1997- author
Title: / Brian Lenahan, Kenna Hughes-Castleberry
Description: First edition. | Toronto: Aquitaine Innovation Advisors, 2023. | Includes bibliographical references and index.
Identifiers: ISBN 978-1-989478-20-2 (paperback) | ISBN 978-1-989478-21-9 (ebook)
Subjects: LCSH: Human-machine systems. | Quantum Computing. | Automation.

Cover design: Brian Lenahan/Kenna Castleberry

Our books may be purchased in bulk for promotional, educational, or business use. Please contact your local bookseller, through www.Amazon.ca or by email at
www.aquitaineinnovationadvisors.com

First Edition: March 2023

10 9 8 7 6 5 4 3 2 1

Table of Contents

Dedication

To the incredible individuals covered in this book: Debbie, Mark, Pantur, Betty, Claude, Sin-Itiro, Satyendra, Elmer, Gottfried, and Muhammad. And to those of us they left behind.

When we fail to look to the past for ideas and inspiration, we fail to leverage one of the strongest creative tools. We call these individuals giants, yet they are human and that for us should be one of the greatest inspirations of all. – *Brian Lenahan*

Books are not written in a vacuum. Thank you to my husband Austen for adding air to my vacuum, and for dealing with my long days of researching, writing, and rambling. And to Robert Charles Wilson, one of the greatest unsung science fiction writers of our time. – *Kenna Hughes-Castleberry*

Acknowledgements

Those who have provided quotes for this book, which we will list below: Dr. Christopher Ferrie, Dr. Scott Aaronson, Dr. Davood N. Rahni, Dr. Brian Evans, Dr. Marina Walt, Dr. Keeper Sharkey, Peter Bordow, Dr. Thomas Ehmer, Dr. Robert Kleinberg, Cedric Wilson, Robert Davis, Abhijit Roy, Mayukh Mukhopadhyay, Santanu Ganguly, Dr. Yidong Chong, Dr. David Thomson, Priscilla Holberton, Phil Holberton, Elisa Torres Durney, Lorraine Tsitsi Majiri, Jason Razien, Elizabeth Reed, Dr. Laurie Calvet, Dr. Federico Rosei, Dr. Hong Tang, Dr. Fengnian Xia, Mohsen Nami, Dr. John Bohn, Dr. Eric Cornell, Dr. Jun Ye, Dr. Brian DeMarco, Dr. Cindy Regal, Dr. Carrie Weidner, Dr. Sara Campbell, Dr. Liz Bridge, and Dr. Mariia Mykhailova.

We also want to thank our many supporters, family, and friends, without whom this book would not be possible. Thank you for encouraging and nurturing our endeavors and following us down many rabbit holes.

Introduction

In writing this, we tried to do things a little bit different than most history of science books. Not only did we pick a cast of characters whose stories had yet to be told, but we also wove these stories into a bigger tapestry of science, showing where many of the foundational ideas within quantum physics, mathematics, and computer science all come from and how they connect.

When Brian came up with the concept for this book, he had been sharing with his many LinkedIn followers the stories of various giants of the past. He quickly found there was a growing audience comprised of thousands of enthusiasts from around the globe. Clearly, for an author, this was enough inspiration to start developing a full book concept.

Given Brian's specific scope of the human narrative for this project, finding a co-writer for the book became an important reality to be able to complete and publish the book in a reasonable amount of time. He decided to collaborate with science journalist Kenna Hughes-Castleberry, as she had a reputation for focusing on the human narratives of stories within her work.

While we could have picked the more well-known scientists like Feynman or Einstein, we believed that there was already enough being said about these scientists that we didn't have really anything new to add. Instead, we focused on the people who have been overshadowed by well-known figures like Einstein and Feynman, such as Sin-Itiro Tomonaga (who won the 1965 Nobel Prize in Physics with Feynman) or Betty Holberton (who worked with Grace Hopper, one of the most well-known female computer scientists). Other stories, like those of Pantur Silaban, Satyendra Bose, or Elmer Imes, have been lost to time or been poorly preserved due to accidents of time or recordkeeping in their countries of origin. Because the world has been oft-dominated by stories of scientists from mainly the same demographic (male and white), we decided that it was time to expand that narrative, and to add a few more names to the list of potential role models for the generations of future researchers. As you'll see, these overshadowed giants are just as important as well-known role models in inspiring and influencing others.

We call these individuals "quantum pioneers" in our book's subtitle, even though not all of them were actual quantum physicists or engineers. Instead, we picked these individuals not only because of their influential discoveries, but because without their work, we would not be able to achieve any level of quantum technology.

Quantum computing, for example, is a rather complicated concept to explain (but don't worry we will explain it), but in simplified terms, it has the potential for faster analysis and computation on a much larger scale, offering significant benefits for researchers, companies, and a plethora of industries. Because quantum computing combines mathematics, computer science, quantum physics (and even other fields like architecture or chemistry), the giants we picked for our book reflect this intersection of fields. This book brings together quantum theorists, experimental physicists, mathematicians, policymakers and more who shaped how quantum has evolved and provided inspiration for generations of future physicists, professors, and enthusiasts.

Stretching over 1,200 years, *On the Shoulders of Giants: 10 Quantum Pioneers of the Past* shares the compelling stories of ten giants of science in a way that makes them more relatable, inspiring, and of course, engaging. Our giants hail from areas of the globe including the Middle East, Indonesia, India, Japan, Germany, and the United States illustrating the breadth of the quantum origin story.

We commence with the illustrious Al-Khwarizmi. Responsible for the development of the concept of algebra (advancing mathematics from geometry and arithmetic to the manipulation of variables and the solving of equations at higher levels of accuracy crucial to quantum computing and atomic/subatomic levels) and the origin of the term "algorithm". Quantum technologies leverage quantum gates whose operations manipulate quantum states and Al-Khwarizmi's discoveries including advancing trigonometry support the precise understanding of these operations. Without these fundamental mathematical ideas, the evolution of quantum computing may have been impossible or at least significantly delayed.

We continue with the German philosopher and mathematician Gottfried Leibniz, the 18th century developer of calculus, binary systems and the compelling concept of monads. The latter idea focused on the precept for quantum computing of "pre-established harmony" (how the universe is composed of indivisible units – monads – the ultimate elements of reality connected in a predetermined, harmonious way). Such connection is similar to the concept of quantum entanglement, the phenomenon where two or more particles are linked in such a way that their properties are correlated, even if separated by large distances; a key concept of quantum computing.

We're also proud to be the first book to have a dedicated section to telling the story of Elmer Imes, the second African American to earn his Ph.D. in physics, make significant contributions to the fields of spectroscopy and quantum physics, as well as engage in the Harlem Renaissance. Not only did Imes bridge the gap between science and art, he also helped fight for equal rights for African Americans and other minorities before the Civil Rights movements of the 1960s. His research helped lay the important foundations for quantum physics and led to important advancements in chemistry as well. As experts believe that quantum computing can significantly improve the fields of drug development and healthcare, these benefits (while yet to be fully realized) echo back to Imes' research.

Next in our selected field of giants, Satyendra Nath Bose, once a young man who reached out to Einstein with a ground-breaking idea, only to work with the quantum icon to revolutionize our understanding of the behaviour of particles at low temperatures. Such behaviour is instrumental to our ability to interact with the smallest of particles by slowing down their movement. Bose's work on Bose-Einstein statistics remains vital to the development of quantum technologies today in performing precise calculations. His work on statistical mechanics helped to inform the design of quantum communication systems and quantum sensing capabilities allowing the more precise measurement of temperature, pressure and magnetic fields.

Sin-Itiro Tomonaga, Japanese physicist and pioneer in the field of quantum electrodynamics (QED), survived the tragedies of World War II, keeping laboratories open and mentoring young physicists. Tomonaga, awarded the Nobel Prize in 1965 for his QED work, commencing in the 1940's explaining the behavior of electrons and other subatomic particles, knowledge fundamental to quantum technologies. Further, Tomonaga's work extended to superposition and entanglement, illustrating that electrons can exist in multiple states at once, and that two or more particles can be linked together such that the state of one particle can affect the other, tracing back today's quantum developments to the giants of the past.

The story of Claude Shannon, considered the "Father of the Information Age," was a natural follow-up to Holberton's own story, as they may have even met (much to our excitement). Shannon's work easily deserves its own chapter, if not its own book, as he created the entire field of information theory. With his research in how information is transmitted digitally (using binary digits or "bits"), Shannon opened the floodgates for new technologies to develop, including quantum technology. Developing quantum technology, researchers lean on Shannon's work on a daily basis, looking at how quantum information can be transmitted and analyzed.

We could not be prouder to include the work of women giants in our book, as their contributions equal those of men in kind if not quantity. Betty Holberton, pioneer computer programmer from the first general-purpose electronic computer ENIAC team, led the development of computer coding which further led to today's quantum algorithms. While quantum computers operate on the principles of quantum mechanics, both require instructions to generate computational results, quantum computers in parallel fashion versus classical computers sequential approach. Algorithms capable of searching through large databases and accurately predicting the trajectory of military projectiles formed parts of Holberton's contribution to the practical uses of computers.

Indonesia's Pantur Silaban, renown for contributions to the fields of general relativity and quantum gravity, illustrated the importance of theoretical quantum physics (TQP) to science. TQP offers quantum experts and enthusiasts the ability to look around the corner (aka possible futures) by thinking outside the proverbial box. When combined with experimental physics, quantum technologies have resulted in the magic Arthur C. Clarke referred to.

Quantum dot technology leaped ahead given the work of Yale Professor Mark Reed, driving smaller size, lower power consumption and high speed in quantum technologies. Efforts towards securing the exchange of encryption keys through QKD (Quantum Key Distribution) and advancing quantum teleportation – transferring information over long distances without physically moving data – trace in significant ways to Mark Reed and his Yale team.

When we think of the quantum field, many often reflect on Einstein, Planck, Bohr, Feynman and many other giants. One prominent physicist's work on ultracold atoms and their application to quantum information science may be less well known but represents critical contributions to the field. Deborah "Debbie" Jin developed new techniques for cooling and trapping atoms to extremely low temperatures enabling the study of quantum phenomena in a controlled environment. We are collectively wiser about the quantum universe because of Jin's body of work.

Why Quantum Histories Are So Important Right Now

Anything with the word "quantum" in it tends to instantly alienate most of the general public, because the word has connotations of mystery or mathematics (Hollywood has not helped with this, as they tend to equate "quantum" with "magic," or "a complicated process we don't have the time or budget to explain"). Unlike fields such as chemistry or biology where you can physically see topics of interest, quantum physics happens at impossibly tiny levels, making it hard to visualize. This leaves us using analogies (such as trampolines or traps) to try to make a clear image of the process for our audience. To try to make it easier to communicate the incredible science and discoveries happening within quantum physics and quantum computing, we've done our best to show the "human side" of this field. Adding a human aspect can not only make this science more relatable, but also inspiring for others.

In conferences, in podcasts, in the media - we consistently hear that quantum technology is years away. So business leaders do not feel compelled to prepare, while pioneers are building out their quantum infrastructure - teams, equipment, software, vendor relationships - as quantum technology incrementally and iteratively improves day-by-day. Showing how far we've come in the past can help us to better predict where the future of this technology might be headed.

Writing About the Quantum Pioneers

Obviously, there are many other scientists we could have added to our list (and there were many candidates who we were disappointed that we could not include), but we wanted to make sure our book was a digestible size for any style of reader. We also found that our lack of physics backgrounds (as Brian is a former banker and currently a strategy leader and Kenna did do some research but now is a physics writer) was helpful in guiding us in how to write about these scientists in ways that were understandable and engaging. Using the theme of "inspiration" to guide us, we worked to create a safe and well-known path through the mysterious dark forest that is quantum physics and come out on the other side. You'll find that these chapters not only discuss the advancements within the multiple fields we mentioned earlier, but also give different perspectives on how scientists actually look or behave. In doing this, we hope that you realize that scientists, like the rest of us, are ultimately human.

By the time you get to the end of the book, you'll hopefully have a better appreciation for the fields of quantum physics, computer science, and mathematics, and see some of the common trends we found while writing this book. Places like the Bell Telephone Laboratory, NBS (the National Bureau of Standards, now known as NIST, the National Institute of Standards and Technology), and the Institute of Advance Study in Princeton all show up multiple times throughout this book. You might also notice, as we did, that there was a significant physics "renaissance" that happened in the early 1900s, with individuals like Einstein or Bohr making an appearance, which inspired many of our giants. While there currently doesn't seem to be another one of these "renaissances" happening for quantum computing, this field is already buzzing with activity and continues to grow as other industries are taking notice of this next-generation technology.

We've also divided the sections of the book into pre and post-World War II. There wasn't a neat line to divide our book into, as many of these stories have overlapping timelines, but we chose the second World War because it not only engaged the entire globe in conflict, but it also sparked many different scientific projects and advancements that a few of our giants participated in.

A Few Stylistic Choices

You'll notice a couple of interesting style choices we made throughout the book. Firstly, we don't refer to ourselves in the first person, instead we say "Brian" or "Kenna," which makes it easier to show who is speaking or commenting than constantly saying "I (Brian)" throughout the text.

Secondly, we call all our giants by their first names. While this might sound off-putting or even disrespectful, we did this on purpose to make them more relatable. For example, instead of "Lenahan or Hughes-Castleberry," we say "Brian or Kenna" to make ourselves connect better to you the reader. In the same way, we do this for our giants (as well as spouses or family members, mainly due to avoiding the confusion of same last names). Instead of "Al-Khwarizmi," we tell Muhammad's story, making him seem more like a next-door neighbor than an ancient mathematician. Obviously, others will continue to be called by their last names to avoid confusion, but this allows us to add another level of engagement and emotional connection to you the reader.

Thirdly, each chapter begins with a short dramatic scene. We did this to show our giant in a historical context. All of the scenes were taken from actual events or retellings but embellished with a bit of poetic license.

As authors we feel fortunate to have studied the scientific wonders represented by these quantum pioneers of the past. There are innumerable quantum pioneers living today who will push forward the techniques and knowledge from our selected group of giants. We enjoyed crafting the stories, and we hope you enjoyed reading them, hope you might even be inspired to create your own quantum journey.

-Brian and Kenna, 2023

GIANTS IN THIS BOOK

GIANT	ERA	ORIGIN
Muhammad ibn al-Khwarizmi	9th century	Central Asia
Gottfried Leibniz	18th century	Europe
Elmer Imes	19th and 20th century	United States
Satyendra Bose	19tth and 20th century	India
Sin-Itiro Tomonaga	20th century	Japan
Claude Shannon	20th century	United States
Betty Holberton	20th century	United States
Pantur Silaban	20th century	Indonesia
Mark Reed	20th and 21st century	United States
Deborah Jin	20th and 21st century	United States

Our Other Candidates That Almost Made the List

NAME	ERA	ORIGIN
Louis De Broglie	19th and 20th century	France
Wolfgang Pauli	19th and 20th century	Europe
Blaise Pascal	17th century	France
Sally Jean Floyd	20th century	United States

SECTION 1 - Pre-World War II Giants

Chapter 1

Muḥammad ibn al-Khwārizmī - "The Father of Algebra"

Pulling the scarf tighter around his face, Muhammad Ebne Moussa Al-Khwārizmī pushes forward as he enters the city of Baghdad. The afternoon sun is sweltering, and the pack on Muhammad's back seems to feel unbearably heavy. He tries not to notice as he focuses on the town around him. Merchants are yelling across the road to each other in a spectrum of languages: Chinese, Syrian, Greek, and even Spanish. Carts, cattle, and people seem to be everywhere. Muhammad finds himself having to weave among the throng. He finally turns a corner and heads down a shaded street, flanked by elaborate domed buildings. There is one that is just up ahead. With intricately carved pillars and lush gardens, it stands out in the middle of the arid city. Muhammad approaches the stone wall and reads the sign: *Bait Al-Hikmah*, "the House of Wisdom." He smiles to himself as he enters the property, sure he will be there for a long time.

***

Called both the "Father of Algebra" and the "Grandfather of Computer Science," Muhammad Ebne Moussa Al-Khwārizmī is probably one of the most influential thinkers of all time. From inventing algebra and algorithms to mapping the longitudes and latitudes of various cities, Muhammad's work shows his remarkable genius. To say he was ahead of his time is an understatement. Yet, it took centuries before western Europe would come to adopt his scientific advancements, mainly due to religious reasons. With his significant impact in many different scientific fields, it's no surprise that Brian and Kenna picked Muhammad as the first "giant" of this book.

Most of the information we could have on Muhammad as an individual has been lost to time. Even his original Arabic manuscripts did not survive, only the Latin versions remain.

Because of this lack of knowledge of Muhammad's life, his story becomes all the more interesting, and even a bit mysterious, as we're forced to use guesswork to fill in the gaps. What is fairly obvious when looking at Muhammad's life, from 795 to 847 C.E., is that he was fortunate to be born during the Golden age of the Islamic Empire. This Golden age was a type of "enlightenment," as scholars were commissioned by Islamic rulers, known as Caliphs, to translate world knowledge into the common languages, as well as develop new philosophies to explain certain aspects of how the world works.

According to Ohio State University professor Dr. Marilyn R. Waldman, the Islamic era began on July 16, 622 C.E. with the *Hijrah* or the migration of Prophet Muhammad from Mecca to Medina. "Between 634 and 870, Islam was transferred from the badge of a small Arab middling class to the domain faith of a vast empire that stretched from the western Mediterranean into Central Asia," Waldman explained in an article for *Britannica*. This expansion caused Arab cultures to mix with indigenous groups, motivating people to become bi-or trilingual. Muhammad Al-Khwārizmī would have no doubt felt this nebulous mix of cultures, and probably wondered, like thousands of others, how to exist in this new "melting pot." By 651, the Islamic Empire was successfully trading with China, adding more cultural variety to the population.

In 740, the city of Baghdad was established as the new capital of the Empire. The caliphs enjoyed all the trappings of court life, including numerous slaves, concubines, and wives. "Europe had nothing to compare with Baghdad," Waldman added. As Baghdad was on the Silk Road, it quickly became a hub for trade and travel. Merchants became wealthy trading exotic goods like paprika or silver from the East for cattle and other goods from the West. Individuals from all over the ancient world including China, India, Africa, and even Spain came to Baghdad. It was *the* place to travel in the 700s and 800s.

Muhammad Al-Khwārizmī wouldn't actually arrive in Baghdad until many years later in his adult life. While many scholars debate when he was born (most claim it in 795 or 788 C.E.), his birthplace may be less of a mystery and perhaps too obvious to those natives of the region. As "Al-Khwārizmī" is a phrase that means "native of Kharazm," many scholars believe he originated there. This name is similar to Leonardo Da Vinci, whereby "Da Vinci," means "of Vinci," no doubt where he was from. Kharazm is a rather unique region as it is "between the eastern Caspian Sea and South Aral Lake in Central Asia," explained Pace University professor Dr. Davood N. Rahni in an interview with Kenna. "Central Asia, from northern India to western China to the Caucasus west of the Caspian Sea, and Asia Minor (today's Turkey) were integral parts of then 'greater Persia' (Iran) both politically and above all, still culturally and Persian as the *lingua franca*. His birth followed the advent of Islam into this region by just less than a decade earlier." Both Rahni and fellow Pace University professor Dr. Brian Evans speculate that Ebne Moussa Khawrazmi may have been of Jewish-Persian heritage. "It is true the Persians had by and large become Muslim through holy *jihad*, but had, and still continue to retain, their historical multiethnic Persian/Iranian identity," they wrote in a 2019 manuscript. Being Iranian himself by heritage, Davood Rahni feels a special connection to Muhammad and his scholarships. "Born and raised in Iran through the completion of my undergraduate chemistry studies, I had become familiar with his myriad contributions to the scientific discoveries and mathematical approach as his trigonometric method of solving a quadratic equation," Rahni told Kenna. "He was then and still now regarded among the nearly dozen 'polymaths' along with Persian luminaries in that era alone as Omer Khayyam, Al-Biruni, Rhazes, Farabi, and Avicenna just to name a few. Each time I heard and still hear the word *algorithm*, the Latinized version of his name that since the past century connotes computing programming, I remain in awe as to how historically revolutionary his approach to a systematic way of 'simplex optimization' and self-consistent method was to tackling some of the most complex real-life challenges still grappled by contemporary scholars. Khawrazmi as they refer to him in the Persianate hemisphere epitomizes such inspirations instilled in me since primary school. In every town in today's Iran, there are schools and universities in most towns named after him."

In navigating both his heritage and the new rule of the Islamic Empire in Kharazm, Muhammad would have no doubt felt confused and unsure of things. To fit into this new society, Muhammad would have been taught Syrian, Sanskrit, Persian, and Arabic. He may also have learned Greek. He lived during the reign of three different Caliphs: al-Maʾmūn, al-Mustaʿṣim, and al-Wāthiq. Al-Maʾmūn would become the most influential Caliph in Muhammad's life, and later his patron. As we have no sources on Muhammad's early life, we can only guess that he continued to study and develop theories for different scientific fields. It is only when he arrives at the House of Wisdom do more facts begin to appear.

The House of Wisdom or *Bait Al-Hikmah* was created by Caliph Al-Maʾmūn supposedly in 832 C.E. With some estimation, Muhammad would have been in his late thirties to mid-forties when the House of Wisdom was established. Its purpose was to be a haven for education and research. With Baghdad's popularity, the House of Wisdom became the center for scientific research and education during the Islamic Empire. It was basically the Library of Alexandria... before the Library of Alexandria. Al-Maʾmūn understood the importance of translation, as it allowed people to learn about new ideas from around the world in their own tongue. As such, he made sure the House of Wisdom had many translators: 47 worked from Greek and Syrian, 17 from Persian, two from Sanskrit, and one from Aramaic. As Muhammad was one of the earliest thinkers to arrive there, he no doubt would have leaned on these translators, especially when developing his own theories.

Key Accomplishments

While Muhammad developed many influential mathematical theories, we'll focus on his other accomplishments first. He made large strides in astronomy by developing the first astronomy "quadrant," which helped determine the time by observing the positions of the Sun and stars. He created the *Zij al-Sindhind* or "Astronomical Tables of Siddhanta," which were solution tables (see chapter 7) for calculating the positions of the Sun, the moon, and various planets. These tables also calculated the dates of future eclipses. While it's easy to think that astrophysicists like Galileo, Brahe, or Kepler were the first to discover planets and their movements, thinkers like Muhammad or even Ptolemy were looking skyward centuries earlier. Muhammad supposedly also wrote works on devices like the astrolabe and the sundial. "Because of his work, sundials were frequently placed on mosques to show the time of prayer," writer Chris Simms stated in a *New Scientist* article. Muhammad is also rumored to have measured the height of the Earth's atmosphere and found the basic concept of a magnifying glass.

He also made large advances in geography. Writing the *Kitāb ṣūrat al-arḍ*, "The Image of the Earth," or "Geography," Muhammad compiled all the current knowledge about the known world into a single manuscript. He oversaw 70 geographers (no doubt at the House of Wisdom) and built off of Ptolemy's earlier work. His final manuscript listed the longitudes and latitudes of over 2,400 cities and natural landmarks, including around the Mediterranean Sea, in Asia, and in Africa. According to Dr. Ibrahim B. Syed, an expert on Muhammad's legacy, a map that Muhammad helped create of the known world shows the Pacific coast of South America, around 600 years before Columbus "discovered" it, which is shocking! Muhammad also measured the Earth's circumference to its most accurate number at the time, using his new creation of algebra.

The word "algebra" is the Latinized word *al-jabr*, which Muhammad defined in his manuscript on algebra: *al-Kitāb al-Mukhtaṣar fī Ḥisāb al-Jabr wal-Muqābalah*, or the "the Compendious Book on Calculation by Completion and Balancing" (quite a mouthful). *Al-jabr* meant "restoration," which for Muhammad meant mathematically adding the same number to both sides of the equation to cancel out their terms. The other key method was *muqabala* or "comparing," or "completion which meant gave the same outcome as *al-jabr* by moving a negative number from one side of the equation to the other and changing its sign." Together these two ideas would help create the fundamental philosophies for algebraic processes. As Muhammad was not using standardized mathematical notation, his equations were in the form of words, such as:

> "If from a square I subtract four of its roots and then take one-third of the remainder, finding this equal to four of

the roots, the square will be 256." Muhammad explains his mathematic process using words, by stating in his manuscript: "Since one-third of the remainder is equal to four roots, one knows that the remainder itself will equal 12 roots. Therefore, add this to the four, giving 16 roots. This (16) is the root of the square. The above can also be stated in terms of modern notation as $1/3(x^2-4x) = 4x$."

Using *al-jabr* and *muqabla*, Muhammad's book showed how to solve polynomial equations using reduction. "In particular, Al-Khwārizmī developed a formula for systematically solving quadratic equations by using completion and balancing to reduce any equation to one that is solvable," explained the Simms in the *New Scientist* article. Muhammad also helped develop the "lattice" or "sieve" method for multiplying big numbers together. Mathematician Fibonacci would later use Muhammad's sieve method and introduce it to Europe, spreading it west. To make his math more exciting and accessible to others, Muhammad showed the many applications it could be used in, including "in cases of inheritance contracts, surveying, tax collection, legacies, partitions, lawsuits, and trade, and in all their dealings with one another, or where the measuring of lands, the digging of canals, geometric computations and other objects of various sorts and kinds are concerned," explained Syed. One specific example in his translated manuscript shows just how much Muhammad was thinking about the everyday uses of algebra.

> "A man is hired to work in a vineyard for 30 days for 10 dollars. He works six days. How much of the agreed price should he receive?' Explanation: 'It is evident that since days are one-fifth of the whole time; and it is also evident that the man should receive pay having the same relation to the agreed price that the time he works bears to the whole time, 30 days. What we have proposed, is explained as follows. The month, i.e., 30 days, represents the measure, and ten represents the price. Six days represents the quantity, and in asking what part of the agreed price is due to the worker you ask about the cost. Therefore, multiply the price 10 by the quantity 6,

which is inversely proportional to it. Divide the product 60 by the measure 30, giving 2 dollars. This will be the cost, and will represent the amount due to the worker.'"

With these many applications, Muhammad was sure he could get more people to adopt his mathematical system. In speaking about his process, Muhammad was recorded as saying: "When I consider what people generally want in calculating, I found that it always is a number."

Developing Some of the First Algorithms

These same ideas helped Muhammad develop the algorithmic process. "Algorithms are a set of rules for solving a problem," Chris Ferrie, a quantum physicist, and best-selling author told Kenna in an interview. "They are important to science and technology because they provide a way to solve problems systematically. Without systemization, attempts to solve problems would be ad-hoc and incomparable." While Muhammad is sometimes credited with inventing algorithms, he actually didn't, as algorithms had been utilized as far back as 2500 B.C.E. by ancient Babylonians. However, Muhammad did help to standardize the algorithmic process and spread it around the world. Algorithms can give many benefits, especially when it comes to making exact measurements. "You don't want to try to go to the Moon, then land thousands of miles away from the Moon, right?" said mathematician Dr. Keeper Sharkey in an interview with Kenna, "I call it the 'exacting craft,' where algorithms help us get to the exact right answer."

Algorithms are especially important in computing, both classical and quantum. "Computer science couldn't exist without algorithms," Dr. Scott Aaronson, professor at the University of Texas at Austin, told Kenna in an interview. "They're also a central part of math. You might as well ask how important numbers or circles are to math." Algorithms are helpful for computing as they give a set of rules for a computer to follow. Whether it's social media or Google searches, algorithms are part of everyday digital interactions. Because they are a part of our daily routines, we probably don't even realize they are there.

They are also important for quantum computing.

Quantum computers work differently than classical computers, as they harness the power of quantum physics to solve nearly impossible problems at faster rates. While a classical computer (like your laptop) runs on "bits," (see chapter 6) using in binary code (see chapter 2) of either 0 or 1, a quantum computer uses the quantum phenomena of "superposition." As Ferrie wrote in his 2023 book *Quantum Bullsh*t:* "quantum superposition is the idea that states-mathematical descriptions-of objects can be added together to obtain other valid states." This offers the chance for faster computing as well as a system to solve insanely difficult mathematical problems. As Ferrie added in his book: "Just before the turn of the twenty-first century, scientists discovered that programs could be created for these quantum computers that solved problems faster and with fewer qubits than bits in a digital computer." There are many different types of quantum computers out there that use different types of qubits including photonic (light particles), superconducting ions, neutral atoms, and others. While quantum computers are still being developed, researchers are developing algorithms to run on these next-generation machines. As Ferrie told Kenna: "Algorithms drive quantum computing research. Quantum algorithms, specifically, show that computers built on the rules of quantum physics are more powerful than digital computers built on the rules of classical physics. If it weren't for the invention of quantum algorithms, quantum computing would not be interesting or even potentially useful. The entire discipline of quantum computing rests on the foundation of quantum algorithms." Brian encounters the importance of algorithms firsthand when he speaks with many of the quantum computing companies in the industry. From his discussions, he's found that one of the biggest challenges they face is building large, practical, scalable quantum computers that perform useful computations while developing algorithms which take advantage of the unique properties of quantum systems.

For a quantum algorithm to be useful, it has to be tried first. "The importance of algorithms in quantum computing is that a lot of them need to be developed and tested and then classified for which type of problem solution they are best suited," Dr. Marina Walt, a Project Manager at the quantum computing company HQS Quantum Simulations said in an interview with Kenna. "This is ongoing work. Of course, to 'justify' quantum computers, the quantum computing algorithms need to be either unique for solving a problem or to beat the classical pendant." Two quantum algorithms that have already passed this test are Grover's algorithm and Shor's algorithm. Grover's algorithm, developed by Lov Grover in 1996, is a general search algorithm for a particular output value. In contrast, Shor's algorithm, developed by Peter Shor in 1995, factors numbers at an exponentially faster rate than a classical algorithm. Both algorithms can be extremely useful in solving some of the most difficult mathematical problems in the world.

But Muhammad had no idea that he would leave this legacy behind when he worked on algorithms. In fact, the word "algorithm" comes from the Latinization of his name: "Al-goritmi." His manuscripts were translated into Latin by Adelard of Bath and Gerard of Cremona in the early half of the 12th century. With this translation, his name would be forever changed. "What is even more perplexing is as if his good name was never carried from the Orient to the West, as rarely his real name is congruent with his Latinized name in the western hemisphere," Rahni said to Kenna.

Besides algorithms and algebra, Muhammad also helped introduce the world to the concepts of 0 and decimal points through common use. "The concept of zero may be one of the most important ideas in mathematics and as related to numerical systems," Nabirahni and Evans wrote. Muhammad used the Arabic-Hindu number system we have today (with symbols 1-10) but also showed that math needed zeros in order to work too. Zeros were helpful placeholders, but could also be used to count more items, as now you could count 100 things instead of just 10. Similarly, decimals were also helpful in breaking up larger numbers, especially for taxation and currency. It would be a few centuries until these notations would be used in Western Europe. In fact, Western Europe was particularly opposed to these new number systems, as they were using Roman numerals instead to count things. There was even a law in 1299 forbidding Hindu numerals to be used in Florence, Italy!

Religious Rage Against Eastern Knowledge

While the Islamic Empire flourished in its Golden age, Western Europe (minus Spain) remained in the Dark Ages. As Syed stated: "To the Muslims, Europe was backward, unorganized, carried no strategic importance, and was essentially irrelevant." Europeans didn't see the Islamic Empire in too favorable of a light either. The Catholic church declared that all Muslims were "infidels" and kicked off the Crusades, as countries tried to assert their Christian dominance. This made it harder for Western Europe to adopt Eastern scientific knowledge and, as Syed posited, possibly expanded the period of the Dark Ages. Those who did travel eastward weren't exactly motivated by curiosity, "The Crusaders were men of action not men of learning," explained expert A. Lewis. When Western Europe did finally accept knowledge from the last, it would help to activate the Scientific Revolution and the Period of Enlightenment (see chapter 2).

With his work on algebra, algorithms, numbers, astronomy, geography, and more, Muhammad Al-Khwārizmī created a legacy that still continues today. He passed away in 847 C.E., and the House of Wisdom declined shortly after due to the political and institutional instability caused by outside forces, mainly the Crusades, the Turks, and the Mongols. While Muhammad had a great reputation in life, in his death he was somewhat immortalized by his work. "I really came to more fully appreciate math and algebra when I first started to learn about his work in college and more so in graduate school," Evans said to Kenna in an interview. "I teach a course in the history of math and I published a book on the subject, and he plays a very big role in both the course and the book. I really enjoy getting students excited about the history of math along with putting a human face on the math they are learning."

Chapter 2

Gottfried Leibniz - "A Man and His Calculating Machine"

The carriage wheels move slowly over the uneven cobblestones of Paris. Each bump makes a concerning sound from the chassis, and the driver hopes that the passenger inside won't complain too much. Contrary to his worries, his passenger Gottfried Leibniz is actually quite comfortable. With his special chair suspended inside the carriage, Leibniz feels only a few bumps, as he continues to scribble notes on his desk. He specially designed the chair as a proto-suspension system so he can write on long carriage rides with little interruption. It is just one of the many things he designs. In thinking about his work, Leibniz looks over at the seat across from him, eyeing a large, ornate box. He sighs and mutters to himself. If all goes well, the mysterious invention inside the box will seal his fate as a leading mind of the 17th-century Enlightenment.

***

When thinking about the process to develop a quantum computer or any sort of next-generation technology, there are a few obvious things you need to know first. One is the notation, which for our current computers, means binary code. Gottfried Leibniz was one of the first thinkers to work on binary notation, helping to create a system to categorize and organize mathematical thoughts and processes. He even published his own manuscripts on integral and differential calculus at the same time as Isaac Newton, but ultimately has not been remembered for this achievement. Instead, Gottfried is remembered for being a philosopher and thinker of the Age of Enlightenment, a period from the late 17th Century to the early 19th Century. Gottfried, as we'll see, was born at the perfect time, when he could explore many different scientific avenues as well as work in the courts of dukes and kings. From his work on binary numbers to his later philosophies that may explain quantum processes like entanglement, Gottfried helped to inspire many different fields of science and technology.

As Gottfried left behind more than 200,000 articles, essays, and pieces of work, his thoughts and legacy have been difficult to study. Many scholars have tried to take on the Herculean task of sorting through Gottfried's possessions, but few know where to begin. While many of the writings overlap, others have no idea how they connect. There are doodles and diagrams, revealing Gottfried's genius while being communicated in almost an entirely different language. This makes Gottfried seem more mysterious as a thinker, and many rumors about him still circulate today.

Born on July 1, 1646, Gottfried Leibniz grew up in Leipzig Germany (back then it was known as Saxony). Both his parents, Fredrick Leibniz and Catharina Schmuck, were reasonably educated for the time, which had a definite impact on his career trajectory. Gottfried's father was a professor of philosophy while his mother's family worked in the book trade. This created an environment where Gottfried was surrounded by books and encouraged to read. As the Gutenberg printing press had been created around 200 years before in 1450, reading was slowly becoming more common, especially among the working class. Thanks to the Age of Enlightenment, a middle class was growing around Europe, interested in learning, via reading, about the world around them. It was no surprise to see young Gottfried take part in this process. He taught himself Latin by age 12 and then moved on to Greek.

The Enlightenment was a distinct period marked by the pursuit of "reason." Philosophers like Diderot and Voltaire began thinking about the condition of humanity, from socialization to government. On the scientific front, the Enlightenment allowed scientists the freedom, and funding, to pursue interesting ideas and theories. Scientists like Copernicus or Galileo could stare into the heavens and find sunspots or elliptical orbits of planets, while others like Robert Hooke could create microscopes to study the world beneath our feet. Thanks to this time period, science and knowledge were growing rapidly, and there was a movement by the working class to be a part of it. Salons and coffee shops around cities like London and Paris centered on intelligent discussions and debates for anyone interested or invited, while the rise of newspapers also spread education to the working class. For Gottfried, the Enlightenment would offer the perfect conditions for him to succeed as a philosopher and thinker.

Unfortunately, Gottfried's father died in 1652, when Gottfried was six-years-old. For the next two years, Gottfried's mother debated letting her son into his father's personal library, as she worried it might be too much for the boy. Finally, at age eight, Gottfried was allowed access to his father's collection of books, many of them about philosophy. As he digested one tome after another, Gottfried began using this new knowledge to puzzle out how the thought process works. As Stephen Wolfram, the famed computer scientist, explained in his 2013 essay *Dropping in on Gottfried Leibniz*: "Even as a teenager, Leibniz seemed to have been interested in systematization and formalization of knowledge."

In 1661, at age 15, Gottfried left home to go to university. In the next year, he finished his bachelor's degrees in both philosophy and law. Gottfried didn't originally begin his career with a focus on science, but instead was drawn into science with his interests in organizing and categorizing knowledge. "There had been a vague idea for a long time--for example in the semi-mystical *Ars Magna* of Ramon Llull from the 1300s--that one might be able to set up some kind of universal system in which all knowledge could be derived from combinations of signs drawn from suitable (as Descartes called it) 'alphabet of human thought'" Wolfram added. Gottfried pursued this idea as he continued his studies, getting a Ph.D. in both philosophy and law. As Wolfram wrote: "He used some basic combinatorial mathematics to count possibilities. He talked about decomposing ideas into single components on which a 'logic of invention' could operate."

During his Ph.D. studies, Gottfried's mother passed away, no doubt leaving him deep in grief. After his graduation, Gottfried rejected a career path in professorship. We don't know why he did this, only that instead he turned to be a court advisor and scholar. Working in the courts of various nobilities gave

Gottfried a variety of tasks to do, from tracking the genealogy of a noble family to negotiating peace talks between territories in Germany and France. His passion for classification and systematization helped him succeed in these many projects and others, like keeping water out of silver mines. While Gottfried tried to use his court positions to try and climb the social ladder, he did not fit in with his peers. He didn't drink or hunt, two popular past times for other courtiers. Instead, Gottfried often found himself alone, tinkering with inventions or thinking of new philosophies. It was during quite a few of these periods that he developed his notations for binary numbers.

Building up the Binary Number System

While a numerical notation system had been around for some time, Gottfried Leibniz is credited with inventing the modern binary system. Before Gottfried, the Hindu-Arabic system (which is still the most used) was developed from the 1st to 4th centuries. This system should be familiar to most of us, as it is a base 10 system and uses multiple symbols for numbers. While Gottfried used this system in his courtier duties, he began looking at new types of systems. He found inspiration in the *I, Ching*, or *the Book of Changes*, an ancient Chinese manuscript dating back to the 3rd century B.C.E. In the *I, Ching*, the reader is introduced to a binary system in yin-yang where yin represents the dark and yang represents the light. Hearing about this binary system, probably from Christian missionaries coming back from China, Gottfried was intrigued to see if the same ideas could be applied to mathematical or logical calculations. Though Gottfried was a philosopher by education, he strived to put his philosophies into practice, including in his development of binary numbers.

"Binary numbers are a way of representing numbers in a place value system with only two digits," explained Dr. Robert Kleinberg, a professor of computer science at Cornell University, to Kenna in an interview. "When expressing binary numbers in writing, the digits are usually represented using the numerals 0 and 1, and the place values corresponding to powers of two arranged in decreasing order as one reads digits from left to right.

Just as we customarily write numbers in base ten using ten digits (0 through 9) with the 'ones' digit on the right, the 'tens' digit to the left of that, and so on, we write binary numbers with their 'ones' digit on the right, their 'twos' digit to the left of that, their 'fours' digit to the left of that, and so on. To give an example, the number 1001101 in binary represents the same number as 77 in decimal, because 77 = 64 + 8 + 4 + 1." Gottfried believed he could build his new binary number system onto a calculating machine, a proto-calculator. Many of these calculating machines already existed in Gottfried's time, including one made by the well-known mathematician Blaise Pascal, who made his own impacts on math and computing. While Gottfried thought about creating a binary number calculating machine, he ultimately abandoned the idea of a base-10 calculating machine.

Binary numbers offer a chance for a place value system, where a digit's position determines the magnitude of the number it represents, as Kleinberg added, "The invention of the place value system was a huge leap forward for mathematics, allowing numbers to be represented using exponentially less space and enabling arithmetic operations to be performed using exponentially fewer steps. Most of the advances in science and engineering in the last thousand years would be hard to imagine without them. The place value system is what allows us to represent the earth's population using a ten-digit number (in decimal notation) rather than using eight billion tally marks."

Place value systems allow flexibility in which number system you use, base 10 or 2 or others. "Babylonian mathematicians used a base-60 number system—which is, incidentally, the reason why we have 60 seconds in a minute and 60 minutes in an hour," Kleinberg said. "Different bases have different advantages. For example, the binary representation of a number is less space-efficient than its decimal counterpart, requiring about three times as many digits. However, learning the multiplication table in binary is almost trivial—you just have to remember that 1x1=1 and that the product of any other two binary digits is zero—whereas it takes quite a bit of effort and time for children to memorize the decimal multiplication table. Binary digits are also easier to distinguish: there's very little danger of mistaking a 0 for a 1 or vice-versa, but a smudge can easily make a 6 look like an 8. It works the same way with electronic representations of symbols: for example, Morse code translates the English alphabet into patterns of binary digits (represented as dots and dashes) rather than attempting to find twenty-six mutually distinguishable signals on a telegraph wire. The ease of distinguishing digits, and the ability to represent arithmetic operations using simple logic circuits, make binary numbers the most favorable representation of numbers on digital computers." The transition from pure mathematical notation to electronics, and even language (like Morse code) shows the importance of the binary number system as a platform to advance, store, or share ideas. Even in developing quantum computers, researchers have used binary numbers to think about the computer readout, as we'll see in later chapters.

Gottfried believed that binary numbers were key to creating a 'universal calculator'' that could do all of the simple calculations in one place. "He wanted to give it a simple 'user interface' where one would, for example, turn a handle one way for multiplication, and the opposite for division," explained Wolfram. Gottfried did create his own calculating machine, albeit not a universal computer, which he showed to mathematician Christiaan Huygens in 1672. He also showed it to British scientist Robert Hooke in 1673, but the machine didn't seem to work effectively. Wolfram describes Gottfried's machine as a "glorified odometer. And just like in Charles Babbage's machines nearly 200 years later, it was mechanically difficult to make many wheels move at once when a cascade of carries occurred." The machine was carried in a decorated wooden box and currently sits in the Leibniz Archive in Leipzig, where it looks like it could still work. Gottfried would spend the rest of his life tinkering with his calculating machine, trying to perfect it.

Leibniz's Stepped Reckoner Machine

Also known as the Leibniz Calculator, the machine was a mechanical calculator invented Leibniz in the late 17th century. The machine was designed to perform basic arithmetic operations, such as addition, subtraction, multiplication, and division, using a system of gears and wheels. At the heart of the machine was a set of stepped cylinders, which were used to represent numbers. Each cylinder had a series of steps, or teeth, that could be engaged or disengaged by a set of levers. The number of teeth engaged on a cylinder represented the value of the number it represented. The machine also had a set of input and output wheels, which were used to input and display the numbers being calculated. The input wheels were used to manually enter the numbers to be calculated, while the output wheels displayed the result of the calculation. The machine was a significant advancement over previous mechanical calculators, which were limited to performing addition and subtraction. Despite its impressive capabilities, the Stepped Reckoner was never mass-produced or widely adopted, due in part to its complexity and high cost, as well as the lack of a market for such a machine at the time. However, the Stepped Reckoner was an important step in the development of mechanical calculators and the history of computing.

While Gottfried never built a working binary calculating machine himself, our modern computers rely on the very system he created. "Binary numbers are the 'natural' language of digital computers, which reads store, process, and display bits of information," explained Chris Ferrie, best-selling author and quantum physicist, in an interview with Kenna. The natural language of binary numbers even gets expanded when it comes to quantum computers. As Ferrie added: "Since we don't speak the 'natural' language of quantum computers-so called 'qubits' we need close analogies. So, we work with bits to label various special qubits of information. If we didn't do this, we would be left with only abstract symbols or large tables of numbers, neither of which are useful to deal with for practical problems." Though classical computers run on the binary system, quantum computers move beyond this system as qubits can be 0, 1, or both. However, to get to a quantum computing system, as Ferrie highlighted, you need to start with a binary system.

Gottfried saw his work in binary numbers as holy, in a certain sense. Quite typical of 17th-century thought, Gottfried believed "binary proves that everything can be made from nothing, with 1 being identified with God and 0 from nothing,"

Wolfram added. Gottfried believed that because his work was "holy" he deserved a medal to commemorate binary numbers. He even drew a design for the medal, including the phrase: "ex nihilo ducendis; sufficeit unum ('everything can be derived from nothing; all that is needed is 1')." Not surprisingly, no medal was made...that is until Wolfram made one for a dear friend. He changed the design slightly but took a copy to the Leibniz Archive to compare his work with Gottfried's drawings. The proposed medal was just one way that Leibniz tried to inflate his status in court. He also wore a ridiculously large wig (which was ridiculed even in his time) to cover up a large cyst on his head. Additionally, he added "von" to his name to become Gottfried von Leibniz, possibly in the German tradition (where 'von' translates to either nobility or simply the word 'from'). While these attempts tried to help Gottfried feel more important, things would only become more irritating with the discovery of calculus.

A Publication Race

In 1673, Gottfried was elected to the Royal Society in London. Only a year later he discovered differential and integral calculus, both extremely useful for many different applications. Instead of publishing his ideas immediately, Gottfried waited 10 years, until 1684 and 1686 to publish his findings on differential and integral calculus respectively. This would prove to be a serious mistake, as Isaac Newton published his own ideas on calculus at the same time. While Newton's marketing strategies were better and he ultimately got the credit, more individuals preferred Gottfried's notation. This debate about who originally discovered calculus continues today, with scholars arguing for both individuals.

While Gottfried no doubt felt disheartened, about losing the "race," he did get a well-paying position in 1692 as the librarian for France's King Louis XIV. This position he kept for the rest of his life, and it allowed him to meet other royal figures, including Czar Peter the Great. This position, like his other courtier ones, allowed him to work on his other philosophies, which many have now interpreted today as early descriptions of quantum mechanics.

The publication race affects not only those within academia, but also in industry. As the head of the Quantum Strategy Institute (QSI), Brian sees this race from an industry perspective, specifically in quantum start-up companies trying to race to file specific patents on new technology. Kenna, in contrast, sees the publication race in the academia sphere because of her work at JILA, a physics research institution that is part University of Colorado Boulder and part NIST (see chapter 10). Because publishing first can affect many different factors, including funding and publicity, the race for publication is present in all fields of quantum science.

Philosophies and Quantum Mechanics

In 1714, two years before he died, Gottfried published the *Monadology*, one of his famous philosophy texts. Many researchers have found sections of the Monadology that could be early descriptions of quantum mechanics. However, it's important to note that Gottfried did not write the *Monadology* with quantum physics in mind, so it's good to be somewhat skeptical of these interpretations.

One of the first interpretations has to do with monads, the main subject of the *Monadology*. While the concept of a "monad" is difficult to grasp, many experts postulate that they are soul-like entities that are indivisible. Gottfried describes monads as "a simple substance, which enters into compounds. By 'simple' [it] is meant without parts." This idea of indivisible, individual properties can easily be found in quantum physics, from string theory to wave-particle duality. In wave-particle duality, a particle can act like a particle in some cases (with specific properties like position and momentum), and in other cases act like a wave (with diffraction or interference). These matter waves are called de Broglie waves, named after another key quantum physicist Louis de Broglie. When combining particle-wave duality with monads, some experts theorize that the monads are similar to elementary particles with different properties, like mass, spin, or charge. As there are different types of monads, this would translate to different types of particles, like bosons or fermions (see chapter 10). Other physicists have thought that the monads could translate to strings within string theory. Yet, monads seem to be such a vague idea that the theory could be applied to almost any small object in quantum physics.

The Principle of Least Action

Gottfried's concept of the "principle of least action" or the principle of stationary action has also influenced quantum mechanics. The principle of least action states that the path of a particle will be the one that requires the least amount of energy to travel or the one that minimizes the action, or the integral of the Lagrangian (a function of the position and velocity of the particle) over time. This principle is closely related to the principle of least time, which states that the path of a particle in a classical system will be the one that minimizes the time taken for the particle to travel from one point to another. The principle says the path of a particle in a quantum system is determined by the wave function of the particle, or a mathematical representation of the probability of the particle being in a particular state at a given time. The wave function is determined by the Schrödinger equation, which describes the dynamics of the particle in the quantum system. This principle is used to explain the behavior of particles in quantum systems, such as the behavior of electrons in atoms and the behavior of photons in optical systems.

Gottfried also discusses the relativity of time and space. In a series of letters written in 1715-1716, at the request of Caroline, the princess of Wales, Gottfried wrote to physicist Samuel Clarke, a friend of Newton's. Some experts theorize that Newton may have helped Clarke write his correspondences, as he and Gottfried debated gravity as well as the nature of space and time. While Clarke (and possibly Newton) saw space and time as absolutes Gottfried argued that they were actually relative. This fell into his principle of the Identity of Indiscernibles, which "states that any entities which are indiscernible with respect to their properties are identical." Basically, if the properties are the same as the subject, they're identical. In Clarke's argument, if space was absolute, then two different locations in space would be indiscernible from each other. Since this is not the case, Gottfried argued that they were both relative ideals. Famed physicist Albert Einstein would later expand on this theory when looking at equations of electricity and magnetism, where motion was only relative, not absolute.

Gottfried may have also been one of the first to describe the idea quantum entanglements where two quantum states become connected and interdependent on each other, though the particles remain individualized (think of a braid in someone's hair, where the braid connects the strands of hair but each strand is still distinct). Entanglement can happen over a large distance, where if one particle is affected, the other entangled particle will also be affected. Entanglement is still of great relevance today, as the 2022 Nobel Prize in Physics was awarded to three physicists who studied quantum entanglement. In the *Monadology*, Gottfried writes: "Wherefore it follows that this inter-communication of things extends to any distance, however great. And consequently, everybody feels the same effect of all that takes place in the universe so that he who sees all might read in each what is happening everywhere." While entanglement wasn't officially proposed until the 1935 publication by Albert Einstein, Boris Poldosky, and Nathan Rosen, Gottfried may have been ahead of the curve in some ideas.

Of course, it depends on your interpretation. It's quite easy to have common philosophies like Gottfried's cause bias in someone's thinking, even if they didn't realize it. As researcher Ludmila Ivancheva states in a paper: "A philosophical work that generates remarkable scientific ideas subsequently 'seep' into modern physical theories and find, or are about to find their experimental confirmation."

Gottfried is not only seen as a thinker but a prominent philosopher. Thomas Ehmer, the Innovation Incubator Lead at Merck Group, a community of science and technology experts, explained in an interview with Kenna that he is particularly interested in Gottfried as a philosopher as Gottfried was able to combine eastern philosophy (like the *I, Ching*) with western philosophy, such as rationalism. Even more modern physicists/philosophers, like David Bohm or David Peat, have expanded on Gottfried's philosophy. Of Bohm, Ehmer said: "Bohm had a very specific view on Quantum Physics, he interprets the potentiality or wavefunction as 'guiding'– and called the bigger concept behind that 'implicate order' – something close to spirituality. He sees the physical world as 'decoding this order.' In the later days, he was fighting with the struggle of language to describe the world – because of emergent behavior, and he was keen on the eastern traditions and also native Indians' world description that is more verb-based than matter (substantive) – something where also Peat was working on."

Both of these physicists/philosophers took inspiration from Gottfried's own writings to further investigate the workings of the world. It was in fact Ehmer who introduced Brian to Gottfried Leibniz originally, a fact that Ehmer is proud of, as it exposed Brian to Gottfried as a potential quantum "giant" (and hopes to do the same with Bohm). Others have also found the crossover between physics and philosophy in Leibniz's case to be quite fascinating. "I think calling out Leibniz's 'Universal Genius'–his holistic and unconstrained views of the intersection of physics and metaphysics almost feel like a 'lost art,'" explained Peter Bordow, a Distinguished Engineer at Wells Fargo. "The application of scientific theories and principles to philosophic questions and puzzles may be the lynchpin in uncovering some of the deepest mysteries being discussed today."

Whether you believe Gottfried did explain quantum physics early on or just was a brilliant philosopher, one thing is clear: he made a significant impact on our society and technology. "He was ahead of his time," explained Ehmer in his interview with Kenna. "This can be a motivation for the quantum community. I think we are also ahead of our time. If we implement quantum computing in a company, maybe others from the future will look back, and wonder about us."

Gottfried's life, and death, were not without mystery. He died on November 14, 1716, in Hanover, Germany. A year before this, Gottfried had been issued a ban from traveling to England. Resources on how or why this ban happened are unclear, but it may have something to do with the fact that a new king, King

George I, had been crowned the same year. King George was originally Georg August, a duke, that Gottfried may have worked for. While records are scarce, it is reasonable to assume that if Gottfried did work for August, there could have been a falling out that led to this 1715 travel ban. After Gottfried died, his papers were sealed, adding further mystery to his status. Wolfram, in his essay, posits that the papers contained state secrets, which had to be locked away for the protection of the state. Of course, the plethora of papers was later unsealed, and have yet to show any sensitive information within them, but due to the sheer number of writings, it's difficult for scholars to find basic information about Gottfried, except for his work on calculus and binary numbers.

Celebrating Leibniz

Germany celebrates Gottfried Leibniz through a combination of academic research, cultural events, and educational programs, highlighting the impact of his work on the scientific, philosophical and mathematical world including the commemoration if his birthday, July 1st, celebrated as Leibniz Day. The Gottfried Wilhelm Leibniz Prize, established in 1985, is awarded annually to outstanding scientists and scholars in Germany. The Leibniz Research Alliance brings together research institutions and scholars to promote Leibniz's work and to encourage further research in his fields of expertise. There are several museums and exhibitions in Germany that focus on Leibniz's life and work, including the Leibniz House in Hanover and the Leibniz Museum in Leipzig. Finally, lectures and symposiums occur in German universities and research institutions regularly on Leibniz's work and its relevance in contemporary research.

In 2013, new research changed the way Gottfried Leibniz's work on binary numbers was seen. Research revealed that populations of Polynesian tribes invented the binary number system decades, if not centuries before Leibniz. While this impactful discovery changed when binary numbers were first invented, it does little to take away from the importance of Leibniz's work in spreading the binary number system around

Europe. “Leibniz is a huge deal,” explained mathematician Dr. Keeper Sharkey in an interview with Kenna. “Without him, calculus, geometric sequences, all sorts of aspects of what it is that we do in science and mathematics wouldn't be here.” When Gottfried died, he was still working on perfecting his ideas. Perhaps if he could see the amazing technology we have today that runs on this system that he helped create, he wouldn’t have worried so much about whether his ideas or machines, were perfect or not.

Chapter 3

Elmer Imes - "A Fine Gentleman"

Jazz drips out of an alto saxophone like molasses as the club's lights dim. A thin young woman takes the stage, her many bracelets clanking as she waves to the crowd. In the dim light, her dress seems almost electric with its many beads.

"My name is Nora Holt, and thank you so much Harlem for having me," she croons before beginning a musical number with the band behind her. As music and cigarette smoke swirl in the air, a woman and a man sitting at a nearby table watch on as a crowd of both African Americans and Caucasians waltz and sway. While both wear the gaudy styles of the 1920s and fit well into the Harlem crowd, the woman seems to have a permanent pout on her face.

"It's not right," the woman huffs, "This isn't a club for mixing." The man's only reply is merely a long exhale of pipe smoke. His eyes remain expressionless behind his round glasses. The woman squirms in her seat, like a chicken ruffling its feathers before preening itself. As she pulls out a faded brass compact, a mousey-looking woman almost collides with their table.

"We have to leave, now," she pants. "It's another raid, the police are always here, those dogs. And you know we are always the ones to get in trouble." Before anyone can say anything, she hops to the next table, spreading the message. The music drones on, as the crowd blissfully continues enjoying the glittering nightlife.

"Figures," the woman clucks, "that's what we get for mixing with these people." She stands up and clicks the compact shut with a snap.

"Hush Nella," the man finally says, before clenching the pipe stem with his teeth. "Let them come. We've done nothing wrong."

***

For most of us, the phrase "Harlem Renaissance" may evoke topics like jazz music or poetry, but probably not science. As one of the few scientists rubbing elbows with Harlem's intellectual elite in the 1920s, physicist Elmer Imes stands out as a curious individual. While Elmer's scientific contributions helped us better understand molecular structures and physics at a quantum level, Elmer also participated heavily in the Harlem Renaissance, hosting many dinner parties and events. His bustling social life not only helped him inspire the next generation of African American physicists, but also led to an illicit affair, a public scandal, and a powerful civil rights case that set the precedent for many other social movements.

His story is difficult to find as information about him is buried or few people have written about it. In fact, this book is the first book to cover Elmer's work directly. But as the second African American to earn a Ph.D. in physics in the U.S. and make huge strides in popularizing quantum theories, Elmer is an obvious choice for a quantum "giant."

Elmer was born on October 12, 1883, in Memphis, Tennessee into a family of privilege. Both of his parents were college-educated, which was a rarity for African American families, as the American civil war had occurred less than 20 years before. While Elmer's father's family were free farmers who fought for the Union army in Pennsylvania, his mother was born into slavery in Mississippi, only to be later freed. Elmer's parents met in seminary school and became home missionaries. As the U.S. was still recovering from the trauma of the civil war, Elmer's parents used their skills to help educate and proselytize former slaves in the southern U.S. Given his parent's professions, Elmer grew up in a home that encouraged learning.

For a long time, Elmer wanted to be a theologian like his father. Elmer did most of his schooling in Alabama. When looking at colleges, he was interested in Fisk University, a private historically black liberal arts college in Nashville, Tennessee. The college was founded in 1866, "primarily for the education of recently freed African Americans," explained physicist and mathematician Ronald E. Mickens in a 2018 article on Elmer in *Physics Today*. Due to its status, Fisk University was one of the best places in the southern U.S. for African Americans to get a college education. This made it harder than normal to get into, so Elmer's father tutored him. With this extra help, Elmer was accepted to Fisk University and completed a Bachelor of Arts in science in 1903. Getting this degree was probably not easy, as Fisk required all its students to do very well in all areas. Elmer studied multiple languages as well as math, science, and art while he was there. The university required such high achievements, because, as Mickens wrote: "The fact that after graduation many students would move to communities in which they might be the only university-educated black person; they would need to provide leadership on issues such as the purchasing and selling of property and guidance on setting up churches, medical clinics, and social services, They also might need to be called onto settle disputes between black people and often- oppressive white communities."

After getting his B.A., Elmer began teaching math and physics at many different schools in Georgia and Alabama, all while still wanting to be a theologian. It wasn't until after his father's death in 1908 that Elmer changed his mind. Being the oldest of three sons, Elmer assumed financial responsibility for his mother after his father's passing. By then his mother was the director of the teacher's residence of the Industrial Missionary Association's School in Beloit, Alabama. She was a very independent woman and would remain so even after Elmer's death. While tending to his mother, Elmer's uncle, who was a well-known physician in Philadelphia, encouraged Elmer to continue pursuing science as a career. This nudge was all Elmer needed to return to Fisk University to get a master's degree in 1915. To make ends meet during his studies, Elmer continued tutoring in science and math, developing the skills he would use later to inspire the next generation of scientists.

Work on Spectroscopy and Quantum Theory

While Elmer tried to get into several Ph.D. programs in the South, he was ultimately rejected due to a high level of segregation. "He lives in a very particular time, during a period of much greater racism and anti-black sentiment in the United States than anything most of us experience today, which is really saying something," said Science Writer and Communications Specialist Robert M. Davis. Davis wrote an influential article on Elmer's scientific achievements in *Symmetry Magazine*, which inspired Kenna to suggest Elmer as a quantum giant. As Davis himself is African American, he found Elmer's story compelling "Elmer was also living during the time of pretty significant change in terms of civil rights, and in the way, African Americans were treated by public institutions," Davis added. Because of segregation and the racist "Jim Crow" laws, Elmer had no choice but to look in the northern U.S. for places to get his Ph.D. He settled at the University of Michigan in Ann Arbor. It would be this decision that would seal his fate as an important scientist in the history of quantum technology.

Due to his degrees being awarded from a black university, the University of Michigan accepted him at first on a probationary study. This was not uncommon for many African American students studying at northern, mostly-white universities at the time. "During the late 19th and early 20th centuries, it was common for northern universities to require black students graduating from black colleges to repeat the last year of the university's undergraduate program because they could not objectively measure the students' qualifications," Mickens wrote. This was no doubt frustrating for Elmer and many other students, as it forced them to add an extra year to their studies. Thankfully, as Mickens added: "black students generally performed extremely well and after receiving their advanced degrees were very successful in their careers." Elmer received a fellowship from 1916 to 1918 that helped relieve the cost of his education. Though Elmer was one of a few African American students at the university, he was well respected by his peers and instructors. This helped him develop naturally genial relationships with white individuals, to the point where he had many as friends later in life. This made Elmer stand out as an individual who was trusting and friendly towards the white race, which was uncommon for African American individuals coming from the southern U.S.

At the University of Michigan, Elmer decided to study under physics professor Harrison Randall, who had only received his Ph.D. about six years earlier. Randall focused on spectroscopy after his own mentor simply handed him a spectrometer and told him to get to work. During Elmer's studies, Randall also worked as a researcher at the National Bureau of Standards (later to become the National Institute of Standards and Technology, NIST).

Spectroscopy sounds intimidating due to its multi-syllabic and mysterious nature, but it's a simple concept. Spectroscopy uses specialized instruments to look at a spectrum, which is a graph or chart that shows the light intensity that is emitted in a specific energy range. In quantum physics, atoms, electrons, and other particles can absorb energy, often in the form of photons or light. When this energy is absorbed, the atom may fluoresce or give off light, as well as jump to a higher energy level. This jump, from the atom's ground state to a higher energy state, can create an energy gap, which can be measured via spectroscopy. Spectroscopy is used in many places in daily life without us even realizing it. From MRI machines detecting tumors to looking at blood stains at a crime scene to even finding spoiled milk, spectroscopy is important in keeping us healthy and safe. Spectroscopy can also be helpful in looking at ice cores from glaciers or water contamination.

Kenna has had firsthand experience with spectroscopy, as she used it when she worked in a plant laboratory as an undergraduate at Colorado State University. As the lab investigated if plants could take in toxic levels of selenium for environmental waste cleanup, the team often tested their own selenium levels to make sure they were at healthy amounts. They did this using hair samples which were analyzed by a spectrometer. The first time Kenna participated in the test, the levels came back in very strange amounts, with high levels of cadmium. It was puzzling until she realized it was due to her red hair dye!

Spectroscopy is also very important for quantum physics and quantum computing. As many quantum computers use qubits (a type of quantum computer bit) that are photonic in nature, a spectrometer is helpful in measuring the processes of the quantum computer. But during Elmer's time, spectroscopy was also important in advancing quantum theory. According to Mickens: "Albert Einstein raised the issue of molecular spectra, which sparked a discussion about methods for quantizing the energy levels of molecules." While scientists during Elmer's time knew that electrons absorbed a certain quantity of energy and became "quantized," they didn't know if this quantization was also happening in molecules or other particles. As Davis explained in his *Symmetry Magazine* article: "Classical physics held that things are generally continuous, rather than quantized. It was just not apparent how useful and widely applicable ideas about quantum mechanics could be."

But Elmer was about to blow this idea right out of the water. Working with Randall, Elmer built an infrared spectrometer, which looks at the infrared spectrum of molecules. As previous spectrometers had poor resolution, Elmer worked to make his machine more precise and with the best resolution possible. In the end, his final apparatus was around five meters (15 feet) long. The device was also quite finicky. As a Ph.D. student in MIT's Ultracold Gases group, Cedric Wilson, elaborated in Davis' article: "'You have to control for temperature; you have to control for everything.'" But Elmer was not deterred, and from his research published two. papers, one in the *Astrophysical Journal* in 1918 and one in *Physical Review* in 1919. While those would be the only papers he would ever publish, they bolstered his reputation as a serious scientist, especially abroad. "He was incredibly diligent and hardworking," Wilson told Kenna in a separate interview. "There are stories of him staying up until 4 a.m. to run his experiment. He did this because during normal hours external vibration from trains going by was gone. He generally went to great lengths to increase the precision of his apparatus by making it larger, while still retaining the precise alignment necessary to make it work as intended." Both of Elmer's papers focused on the spectroscopy of the gases hydrogen bromide (HBr), hydrogen chloride (HCI), and Hydrogen fluoride (HF).

The results of Elmer's work showed multiple thick absorption bands at specific intervals. From these bands, Elmer could begin to determine the structure of the molecules. This is because "whether a molecule absorbs a certain wavelength of light is related to the structure and mechanics of the molecule, along with its rotational energy," wrote Davis. The molecule, due to its structure, will only absorb a specific wavelength of light, causing its energy to become quantized. Classical physics didn't predict this process because it theorized that energy was continuous, not in discrete amounts. In his 2023 book *Quantum Bullsh*t,* quantum physicist Chris Ferrie explains the important contrast in energy definitions by writing: "In classical physics, energy is continuous (smooth). In quantum physics, energy is discrete (chunky)." Thanks to Elmer's work, quantum theory expanded to molecules and atoms instead of just electrons. "Thus Imes's work formed a turning point in the scientific thinking, making it clear the quantum theory was not just a novelty, useful in limited fields of physics, but of widespread and general application," stated physicist and spectroscopy inventor Earle Plyer about Elmer at a later date. With Elmer's data quantum mechanics became connected to atomic physics, showing that quantum worked on nearly everything. His spectroscopy results also found that there were clear distances between the atoms and the molecules, unveiling more detail about their structure.

Elmer's work on measurement precision in quantum aligns closely to areas beyond spectroscopy, specifically quantum sensing. One of the most promising areas of quantum sensing is in the field of magnetometry, where the goal is to measure extremely small magnetic fields with high precision. Within magnetometry there is, "quantum magnetometry," which uses the spin states of individual atoms or electron spins to measure magnetic fields with high precision. Another area of quantum sensing that is gaining attention is the field of quantum imaging, where the goal is to use quantum systems to create images with a high degree of resolution and sensitivity.

Elmer's two papers would later be added to physics textbooks and would be instrumental in molecular analysis and even things like drug discovery. As quantum computing offers more efficient ways to simulate molecular interactions and eliminate ineffective drug options, Elmer's work was the first to show the power of quantum physics on a molecular and atomic level. As Wilson told Kenna: "Quantum mechanics tells us how very tiny things work, but all theories need to be supported by experimental evidence, and his experiment fit a missing piece into the puzzle. At the time, quantum predicted that molecules would have spectroscopic features that looked like a fine toothcomb, but experiments before Imes only saw large blobs that were the outline of the comb shape (as if all the bristles were blended together). Imes built a cutting-edge experiment that was able to see the teeth in the comb. I do think his work should be more well known—most folks in my field don't know his name or his contribution to our understanding of molecules."

These two papers helped solidify Elmer's reputation as a spectroscopy expert, especially from abroad. Later colleagues and visiting scientists from Europe would comment to Elmer's students that they had no idea he was African American. Thankfully, this new revelation did little to change their working relationships. Even right after publication, Elmer became a sensation. According to Mickens: "German theorist Adolf Kratzer quickly demonstrated the importance of Imes's work for the newly emerging quantum theory."

While these papers bolstered Elmer's reputation, they did little to help him find a job in academia. After World War I, just as Elmer was graduating from his Ph.D. program, a large number of African Americans migrated from the southern U.S. to northern cities, worrying many local white populations. With tensions on the rise, many northern institutions were already wary of hiring African Americans for teaching positions. Swimming in these turbulent waters, Elmer struggled to convince any university to hire him as a professor. After enough rejection letters, Elmer decided to try his luck in the scientific industry. He would work in this corporate world for the next eight years, consulting at companies like Burrows Magnetic Equipment Corporation or Everett Signal Supplies.

In industry, Elmer was able to successfully file four patents, all for devices measuring the magnetic and electrical properties of materials and the measuring protocols themselves. While we don't have many more details about these devices, we do know that Elmer was one of the few African Americans at the time to receive a patent. Unfortunately, his patents caused him future troubles, as they prevented him from further publishing his research, and from immediately taking up a teaching position later in 1930. They also provided little financial reward. One good thing did come out of Elmer's industry stint, he met his future wife Nella Larsen, and became part of the Harlem Renaissance crowd.

The Harlem Years

Nella Larsen has only recently become "rediscovered" as an influential Harlem Renaissance writer and poet. Though Larsen was African American, she did have white family members, and through her parent's divorce and remarriage she became alienated from her white family. This no doubt caused her to have conflicted feelings toward white individuals, as well as struggle with the idea of abandonment. Elmer met Larsen while working with the organization the Circle for Negro War Relief in New York City. Larsen also worked for the organization as a nurse. According to Larsen's biography, Elmer was "of medium height and slender build, about half a foot taller than Larsen and eight years older... he was good-looking if not classically handsome." Larsen and Elmer hit it off almost immediately, as they both were intellectually driven and enjoyed pushing the boundaries. After some dating, the two were married by Elmer's theologian brother William on May 3, 1919, in New York City.

From there the couple moved with Elmer's jobs, first to Jersey City, New Jersey, and then to Harlem, New York in 1927. "I remember getting chills when reading about Harlem in [Larsen's] biography because I've lived in New York for 15 years and so much of the life that was described just sounded so familiar," Davis told Kenna in a later interview. "It was dinner party after dinner party, and running around town going to see shows, and gossiping with your friends, and friends being in marital arguments. It was about who was cool and who had something to say. While there was nothing in the biography that indicated Imes was someone with his own creative ambitions, he was definitely someone who appreciated and supported the arts, and respected his wife's work as an artist." Harlem was an area of the city that attracted the intellectual, artistic, and social elite. In Harlem, Elmer and Larsen became friends with such famous individuals as sociologist and activist W.E.B. Du Bois and poet Langston Hughes. Larsen took to the culture right away, abandoning nursing in 1921 to work as a librarian at the Harlem branch of the New York Public Library. This position gave her the time and flexibility to write as one of the many voices of the Harlem Renaissance. "The basic premise of the movement was that cultural productions would prove the humanity of black communities, and, consequently, allow black people to achieve political, social, and economic equality in the U.S," explained Mickens.

To fully enjoy the opulence of the Harlem Renaissance, Elmer and Larsen would often attend jazz clubs or meetings of the NAACP (the National Association for the Advancement of Colored People). Some of these clubs and meetings allowed white attendees, which many in Harlem frowned upon. Police often raided these "mixed" or "black-and-tan" establishments more than others to try to prevent any integration. Some clubs even barred one race or the other entirely, and because of its high-brow reputation, the Harlem Renaissance banned many whites from African American establishments, as ironic as that sounds. While Larsen gave no direct indication that she disproved of this integration, the characters in her books are all against it. Elmer, on the other hand, didn't care about integration, as he was used to socializing with both races. He just enjoyed the atmosphere of the Renaissance itself. "It is cool to think that 50 years before the Civil Rights Act, before so many things I take for granted; that I can live as someone not encumbered by segregation, there was not just one man, but a whole community of people living in a center of art and intelligence," Davis commented in his interview. "Elmer was definitely a product of enormous privilege. He came from a pretty wealthy black family. But in a lot of ways, these people were just like us. These weren't a bunch of dowdy old conservatives. They were smart young people who were subversive and rebellious. These were real human beings who raged against the oppressive systems around them, just as much as we do today."

Unfortunately, the Great Depression put an end to the glittering Harlem Renaissance. Once more out of a job, Elmer began looking again at teaching positions. After some negotiations, he received a professorship at Fisk University, overseeing the physics department, which was mainly just him. Larsen did not want to travel to Nashville as her community was based in New York. "Selfish little beast," Elmer called Larsen for her stubbornness, and in frustration moved without her as Larsen stayed behind. The move would prove to be the beginning of the end of their marriage.

A University Scandal

Instead of buying his own home in Nashville, Elmer began staying with his mistress Ethel Gilbert. He had started an affair with Gilbert several years before. Gilbert worked as a fundraiser at Fisk. She started in 1928 when the University was in debt of $110,000, around $1.9 billion in today's economy. This cost was devastating for the University, but Gilbert had a plan. With her expertise, Gilbert cleaned up the debt and even used her networking skills to secure a multi-million-dollar endowment. This allowed Fisk to hire the best of the best, like Elmer. Gilbert was infatuated with Elmer, stating that "the night after he left was like the parlor grate with the fire gone out." Oh, and Gilbert was white.

Because of the interracial part of their relationship, Elmer and Gilbert tried at first to keep it a secret. It would later cause a huge scandal for Fisk, and cost Elmer his reputation. The publicity could have even cost Elmer his life, as lynchings in Tennessee were still occurring. According to the Tennessee Encyclopedia, around 2,805 people were lynched between 1882 and 1930 in Tennessee, with around one victim a week. As Emmett Till's case, probably the most well-known lynching incident due to accounts that Till whistled at a white woman in Mississippi, wouldn't happen until 1955, an African American man could be lynched for doing almost anything. That's what made Elmer's affair not only illicit but dangerous.

Though Larsen knew about the affair, she kept it to herself, for fear of losing her marriage. She still lived in New York, though Fisk University offered to build a house for Elmer and her to keep up appearances. The house Elmer helped design would become an English-style cottage with separate areas that he and Larsen could live in. Before Larsen moved in, she traveled around southern Europe on a Guggenheim Fellowship, becoming the first African American woman to win one. Before she had left, she had agreed to an "open" marriage with Elmer. While Larsen was away, Elmer spent more time with Gilbert but also worked on building up his physics department.

As the Chair of the physics department, Elmer had free reign. He created a curriculum for undergraduates with a focus on spectroscopy and research. He helped design a new science building for the University and participated in many clubs and school events. Inspired by the Harlem Renaissance, Elmer also created a "cultural physics" class for students and even colleagues. The class focused on the history and impacts of science and society.

Unfortunately, Elmer's domestic life was about to put his research on hold. In something like a soap opera, Elmer found out that Gilbert was pregnant around the same time that Larsen finally told him she knew about his affair. They both finally agreed to the divorce, on the condition that Larsen site "cruelty" as the reason instead of adultery, so as not to avoid Elmer losing his job, or creating a scandal. Elmer wrote during this time that: "If I am to work and do the things I must do for myself, there can be no publicity-my job would be gone in ten minutes and I wouldn't be able to do anything for anybody." In the terms of the divorce, Elmer was to pay Larsen a monthly stipend, give her all the house furniture, and pay her legal fees. He never financially recovered after the divorce and married Gilbert only a week later.

A scandal did break out about the affair and the divorce, as a 1933 edition of the *Baltimore Afro-American* newspaper read "Imes Must Pay." Surprisingly, little changed on campus. Gilbert was forced to resign and later tried to get her job back. Many had actually supported the affair, as Elmer and Gilbert were often seen together in public and both were loved by the students.

After the divorce, Larsen moved back to New York and Elmer moved his mother into Gilbert's old apartment.

Elmer continued to mentor students while building Fisk's physics department. He joined the American Physics Society (APS) and the American Institute for Electrical Engineers, though he refused to attend segregated meetings in the southern states. Elmer's life ended painfully on September 11, 1941, after metastasized throat cancer caused him hours of suffering. His ashes were scattered around the house he built for Larsen and himself. He left the bulk of his estate to Ethel Gilbert. In an obituary written in *Science*, Elmer's laboratory was described as a "mecca for those who sought an atmosphere of calm and contentment. Peacefully smoking his pipe, Imes could always be relied upon to bring any discussion an atmosphere of philosophic soundness and level-headed practicalness... In his passing, his many friends mourn the loss of a distinguished scholar and a fine gentleman."

The Civil Rights Activist

While Elmer made many influential scientific discoveries and mentored many, he also was passionate about civil rights. Many more African Americans could have become academics and researchers like Elmer, but, as Mickens posited, they instead became civil rights activists. Elmer was able to be an activist while doing science. "Just the fact of being a black person in a space that is typically not open to you is activism," Davis said in his interview. "The world has changed where more of us have the privilege of where and when we are activists. We may be oppressed by systems, but we aren't always proactively involved in the fight against those systems. But as you look further and further back in history, you didn't really have a choice."

A car accident in 1931 caused Elmer to really feel this lack of choice. Juliette Derricotte, the Dean of Women at Fisk University, and three African American students were seriously injured in a car accident in Dalton, Georgia. They were taken to a hospital but turned away because it was a "whites-only" establishment. The group then went to a Mrs. Wilson's house, who took care of the four individuals as best she could, but she was not a medical expert. Due to a lack of medical care, Derricotte and one of the students died the day after the accident. Elmer was incensed and urged the NAACP to open an investigation into what happened. He also wrote to Du Bois about the event, who in turn wrote about its effects on society. Protests occurred across the country after the accident, especially at Fisk University, yet little legal action was actually taken. Elmer continued to work with the NAACP and other groups to improve equality for African Americans. "He had every opportunity to live in the ivory tower and not get involved," Davis added in the interview. "But he came down and got his hands dirty."

Elmer's story is important on multiple levels and conveys the importance of resilience. "It feels a little hackneyed to say sometimes but representation really does matter," Davis stated. "Seeing examples of people doing things when they look like you helps you to see that you can achieve those things too. It really is as simple as that. But even though I feel like I've been hearing about the importance of representation my whole life, it isn't safe to assume those sorts of messages are penetrating the physics community. I think it's really exciting to see some historical figure you strongly identify with, and then learn that he actually founded a physics program that graduated hundreds of black scientists and educated even more young black people in physics fundamentals." To honor Elmer's work, the National Conference of Black Physics students bestows an annual Elmer Imes award. For Davis, Elmer's story is one that needs to be told more, as he explained that "It's a reminder that the little things you do in your life can reverberate into the future in a really meaningful way."

Chapter 4:

Satyendra Nath Bose - "The Comet"

A resounding horn is heard around the bay as people scurry across the large steamer like ants on a rotting log. Amid passengers shouting and porters stumbling to carry luggage stands one man: Satyendra Nath Bose. His tall figure leans over the railing, and from the peaceful look on his face, he seems to not even notice the commotion happening behind him.

"Sir, we're going to set off momentarily," a voice calls to him. Satyendra turns to see a small teenage deckhand looking up at him quizzically, Satyendra smiles and nods. He folds the small card-like piece of paper and puts it in his pocket, patting it.

"Is that your passport?" The boy asks.

"No," Satyendra responds, "It's something much more important."

***

The story of Satyendra Nath (also written as Satyendranath) Bose has been interpreted by historians in many ways over the decades, and yet has not become part of mainstream knowledge. Most individuals, if asked, will most likely not know who Satyendra is, that he's Indian (many people assume he's actually German and his name is just misspelled from Böse) and that bosons, one of the two classes of fundamental particles are named after him. While Satyendra may currently remain out of the public eye, his story would have been further lost to time if it wasn't for the help of Albert Einstein. Einstein's relationship with Satyendra, as we'll see, was mainly one-sided, and could have been the driving factor for why Satyendra seemingly only really made one important contribution to quantum physics before seeming to leave the field entirely.

It's important to note that Satyendra's story (along with the stories of other Eastern scientists in this book) is somewhat limited. Because he grew up in a developing country, and a British colony, many of the early records we would have had on his life have been lost. Others have had to be translated, causing differences in interpretations. Because of these differing interpretations, and his close historical associations with Einstein, many get upset that Satyendra is not just seen by himself with his own work. Ultimately one needs to realize that Satyendra's story continues to be told in part by people who are not of his ethnic background and that can give different readings of Satyendra's accomplishments. Being cognizant of this, we will do our best to present the facts only, and give a balanced view of how Satyendra's story has been analyzed.

Many historians interpret his story as one of being in the right place at the right time; where he was able to lean on Einstein's generosity to expand his own career. Others read Satyendra's contributions as a "shot in the dark," and that this brief stroke of genius made him a "flash in the pan," burning brilliance once and then never again. Still, even other experts believe that Satyendra's genius was worthy of even more credit than he was given and that it should have earned him a Nobel prize. One 2012 article by Dr. Deepanwita Dasgupta, philosopher, and professor at the University of Texas at El Paso remarked: "Bose had done something, which, to put it bluntly, could not have been done: he created a fundamental concept in quantum mechanics out of thin air before quantum mechanics was officially born." No matter how you interpret Satyendra's life, it's clear that he was absolutely brilliant. Rumors about his genius continued beyond his death, as people heard stories about Ph.D. students getting their degrees based on ripped-up paper scraps found in Satyendra's wastebasket (this rumor might actually be true, as one colleague of Satyendra's once said: "Dr. Bose finds pleasure in complex problems. His enthusiasm dies once he solves them. He throws the proofs into the wastepaper basket, never bothering to send them to any journal."). Another story described how Satyendra immediately solved a difficult equation for Niels Bohr during a lecture in Germany. While we will never know if these rumors are true, they do depict Satyendra's amazing abilities as a scientist and a teacher to help encourage learning among his peers and students.

Telling Satyendra's story is also telling the story of Indian nationalism and India's independence in 1947. Satyendra played an important role in helping his country gain its independence, just not in the way you might think. With the combination of his brilliant work and his support of local communities and students around him, Satyendra Bose stands out as a powerful role model and a unique individual in the history of quantum science.

Born January 1, 1894, in Calcutta (now known as Kolkata) India, Satyendra entered a world of exploration and imperialism. His family belonged to the Kayastha community, which historically was a caste of record keepers and scribes. As the oldest of seven children and the only son of the family, Satyendra knew he had many duties to follow. His father, Surendra Nath Bose, was an accountant and later became more of an entrepreneur, setting up his own chemical and pharmaceutical company. When Satyendra was young, his father would scribble mathematical problems on the veranda floor for his son to solve while he was away at work. Satyendra would solve all his father's problems and proudly show him when he returned. Thanks to his father's encouragement, Satyendra became confident in his mathematical abilities at a young age.

Growing mathematical confidence in young children is crucial for parents to work on, as studies show that being more confident in math is linked to higher self-esteem. A 2017 survey by Microsoft showed that girls originally become interested in STEM (Science Technology Engineering and Mathematics) subjects at age 11, but then quickly lose interest by age 15. The survey suggests that this process could be due to the gender stereotypes and social expectations that young girls are faced with by the time they reach age 15. While the survey does not mention self-esteem becoming a factor, the results from other studies suggest that this is another influence to consider. For parents who want their children to have both high self-esteem and mathematical confidence, it's important to show mathematical problems in exciting contexts (like building with Legos) and presenting problems often. Just as Satyendra's father helped him to grow in this confidence by constantly giving him problems, so too can parents do this today. In fact, it's easy to see that Satyendra also had a healthy level of self-esteem throughout his whole life, partially due to his early education.

With his father's guidance, Satyendra began his schooling at age five. Though he reportedly had weak eyesight (and would later wear glasses), he continually received some of the highest marks in his class (and would set college records for his grades that have yet to be broken). At just 15 years old, Satyendra entered Presidency College. He would receive both his bachelor's and master's degrees in 1913 and 1915, respectively. Both of his degrees were in mixed mathematics, which was "applied mathematics dealing with astronomy, dynamics, hydrodynamics, and certain mathematical topics," according to a 1975 Royal Society Memoir on Satyendra. Having a wider range of mathematical applications under his belt would help Satyendra immensely when he began studying quantum physics.

In 1915, during his last year of schooling, 20-year-old Satyendra married 11-year-old Ushabati Ghosh, the daughter of a well-known Calcutta doctor. While we don't know the full context of this marriage, it would be safe to assume that this marriage was probably arranged. *According to Satyendra Nath Bose-His Life and Times: Selected Works* by Kameshuar Wali: "Bose was not keen on getting married so early, but he could not go against his mother's wishes." Whether or not the marriage was arranged, both Satyendra and Ushabati did their best to make it work. They would have nine children together, though only seven survived childhood. Satyendra did his best to take care of his family. As Wali wrote: "Unlike traditional husbands, he was very thoughtful of her in small ways. 'If she slept late, Satyendranath would take care of arranging breakfast for the children, prepare for himself a couple of toasts and tea, and then run off to college.'" While the marriage may have been beneficial for Satyendra's family, it would become a challenge for Satyendra's career, as it limited the types of funding and jobs he could receive.

India's Imperialist Uncertainty

During Satyendra's lifetime, India was undergoing revolutionary change. Great Britain's rule over India began in

1763 after the British defeated the French in the Seven Years' War. Using the British East India Company as a ruling body, Great Britain controlled the new colony with a rigid social structure. In 1857, a rebellion, called the Sepoy Mutiny, caused the British to rethink how they controlled the vast expanse of India. This created the period of the British raj, where Great Britain had more direct rule over the country until India's and Pakistan's independence in 1947. The raj was handled by a viceroy or a lord that stood in for Queen Victoria. There were several viceroys, but one of the ones who caused the greatest upheaval was Lord Curzon.

In 1905, Lord Curzon decided to partition Satyendra's homeland of Bengal, and rearrange its borders. According to Wali: "By partitioning Bengal, Lord Curzon, one of the most imperialist reactionary viceroys of British India, had intended to curb Bengal's influence in the rising national struggle for independence. Instead, this created a furor against the Raj and initiated a strong *Swadeshi* movement of self-sufficiency, symbolized by the boycott of foreign goods, a public bonfire of British mill-textiles, reliance on homespun industries, and street gatherings singing patriotic songs." Satyendra was a teenager during this uncertain time and the revolution made quite an impression on him. "Among the young, it generated a strong national awareness and dedication to the service of the country," Wali added. "Talented students like Satyendra Nath thought science, engineering, and technology were the ways to lead the country to progress and prosperity." Because of this, Satyendra and his college peers, including scientist Meghnad Saha, worked hand to show that India had just as many brilliant thinkers as the West. In a 2022 article, Dr. Partha Ghose, a Distinguished Fellow at the Tagore Centre for Natural Sciences and Philosophy in Kolkata explained that: "a galaxy of brilliant and rebellious men appeared on the scene, embracing Western learning and science from essentially an Indian standpoint." This idea would stay with Satyendra for the rest of his life, and he would return to it decades later.

While Satyendra worked to support the *Swadeshi* movement in the academic arena, he also worked to support it on the local level. He was limited in his support, however, as his father didn't want him to do anything that would jeopardize his career. "As the eldest and only son, he was made aware of his responsibilities to his family," Wali wrote. "He obeyed his father but remained a strong supporter of those who participated in the *Swadeshi* movement, helping in whatever way he could, acting as a courier, providing shelter to fugitives from police pursuit, or helping out with money." He also helped teach many of the protestors at the Working Men's Institute during the evenings. Satyendra would continue to help his country as it struggled and finally gained its independence in 1947. Satyendra even spoke to Einstein about India's revolution in 1925 when he visited Berlin. As he wrote in his journal (and recounted by Wali):

> "He [Einstein] confided to me: 'I think Englishmen are better than other Western colonial nations, and I feel that they are by far better than the French and the Dutch. You should not be surprised at a German like me praising the English (after the World War). Now, tell me do you really want that the British should quit your country?' I said, 'Of course, we all want to determine our own destinies.' He was not quite convinced. He raised a hypothetical question. He said, 'Suppose, there was a button near you and all the Englishmen would quit India if you were to press it. So, would you press that button?' I smiled and said, 'If God were to grant me such an opportunity, I would not hesitate even for a moment to press it.' 'Really,' he said and kept quiet for a while."

A Transition into Quantum Physics

After receiving his master's degree in 1915, Satyendra felt stuck. Because of World War I, India was cut off from most of the West, and therefore Satyendra and his fellow graduates had very little knowledge about other scientific advancements across the world. Satyendra tried to pursue a Ph.D. but was ultimately rebuked by a professor who told him he just wasn't cut out for research. According to data consultant Mayukh Mukhopadhyay in an interview with Kenna: "Even being the top of his batch in Physics Masters, he delisted himself from Ph.D., due to his outspoken nature towards his supervisor Prof Ganesh Prasad. Prof Prasad papers were extremely low scoring to otherwise brilliant candidates of the Physics department and SN Bose out spoke his mind, even with his good standing in marks, that it was not the pupil but the teacher himself, the root of the problem of the mark." After quitting his Ph.D., Satyendra began private tutoring and found that he enjoyed teaching students about math and science. However, it was still difficult to find a stable position, which was becoming a bigger need as Satyendra's family continued to grow.

Thankfully, an opportunity presented itself in 1916, when Sir Asutosh Mookerjee, who was the Vice-Chancellor of Calcutta University and a judge of the high court, established the University College of Science, which would teach advanced science courses to students. With his influence, Mookerjee was able to make the institution more autonomous and freer from the British. This meant, for the first time, Indian professors would be needed to teach these advanced courses instead of their British counterparts. Both Satyendra and Saha approached Mookerjee to ask for positions in the new institute. After some discussion, Mookerjee assigned them to both learn advanced physics (such as quantum physics and the newly discovered Theory of Relativity) during a probationary period, before being able to teach. Both men eagerly agreed and were excited about their new careers.

There was just one problem: a lack of educational resources. Because of India's isolation during World War I, there were not many books or experts within India that could help teach Satyendra about advanced physics. After asking all his connections, Satyendra finally struck gold when he discovered that professor P.J. Prühl, a physics teacher at Bengal Engineering College, had several books on advanced physics, including one by Maxwell Planck. The only downside was that the books were written in German. While Satyendra's colleague Saha was fluent in German and had no trouble reading the texts, Satyendra had to learn both German and advanced physics simultaneously. As Ghose explained: "Saha was the first to apply the Bohr theory of atoms to calculate the temperature, pressure, and chemical composition of stars, from the degree of ionization of the atoms near their surface." Thanks to his brilliance and help from Saha and others, Satyendra was able to accomplish this in a relatively short amount of time.

In 1917, Satyendra and Saha were able to finally teach at the University College of Sciences. A year later they published a joint paper on molecular statistics. While teaching, both Satyendra and Saha also worked on translating Einstein's various publications into English. They compiled their work into a book, which then became the subject of a legal dispute with other researchers over who had permission to publish the book. To settle the dispute, Satyendra and Saha reached out to Einstein directly. Satyendra later recounted that "Einstein was very generous. He said that as long as the book remained in circulation inside India he had no objection," thus allowing Satyendra and Saha to win the dispute. While objectively the 1919 book wasn't a very good translation of Einstein's work, it did help to make it more accessible to a wider audience. This would not be the last time Satyendra would lean on Einstein's generosity.

Working on Quantum Statistics

In 1921, Satyendra decided that the University College of Sciences was no longer a good fit and instead moved to the University of Dhaka (also spelled Dacca) as a Reader (a position below a professor; this was because he did not have a Ph.D. so he could not be a full professor). As Satyendra stated later: "The University College of Sciences in Calcutta was beginning to be rather crowded, and I thought that Dacca would offer a freer opportunity to carry on studies in all kinds of subjects." At Dhaka, Satyendra began to teach advanced physics, including Planck's formula. According to some sources, it would be during a lecture that Satyendra would discover his own derivation of Planck's formula that would change physics forever. Other historians called Satyendra's discovery a "lucky mistake." However, in looking at Satyendra's mathematics background and his drive for continuously learning, a "mistake" does not seem like the right descriptive word for what he did, as he applied his mathematical thinking to a complicated problem.

To understand Satyendra's contributions to physics, we must go back a bit. In 1900, physicist Maxwell Planck was working to solve the problem of black-body radiation. Black bodies, in physics, absorb and then emit all the radiation they are exposed to. While this might seem like a straightforward idea, explaining the physics behind it was proving to be a puzzle for physicists in the 1900s, as the experiments on black-body radiation did not agree with previous theories. According to Dasgupta: "Experimental work on black bodies was sufficiently advanced during this time, making the mismatch quite obvious." To try to fix the mismatch, Planck tried a rather weird assumption about the walls of the cavity, or box, where the black-body radiation was happening. "Let's assume, he said, that the...walls do not radiate energy continuously as demanded by the classical theory," Dasgupta wrote. "Let us assume instead that the energy is radiated by the cavity wall in small, discontinuous jumps." This assumption led to Planck's law, a mathematical equation that helped correct the mismatch. Even though he now had an agreement between experimental results and mathematical theory, Planck still didn't know why this worked. Other physicists worked to answer the question of "why" including Einstein. Einstein did expand on Planck's work by showing that the theory worked if the gases were seen as individual particles and not waves. This still didn't answer *why* the mathematical theory worked though.

When Satyendra began studying both Planck's and Einstein's calculations, he looked at them from more of a statistical perspective. As Dasgupta explained: "The main problem that his predecessors had left him with—in spite of their many elegant constructions—was that these different strategies failed to add up to any coherent mathematical framework…What Bose's contribution did was to point out that this structure rests on a fundamental assumption-that each particle in that gas maintains its distinguishable character from all the others." With this expansion on Einstein's work (the gas was made of particles and now the particles were distinct), Satyendra had helped to answer the "why" question for the mathematical explanation of black-body radiation. When speaking about his discovery, Satyendra said: "As a teacher who had to make these things clear to his students, I was aware of the conflicts involved. I wanted to know how to grapple with the difficulty in *my own way.* It was not some teacher who asked me to go and solve this problem. *I wanted to know."* In 1923, Satyendra documented his discovery in a short, five-page paper (this would not be the winner for the shortest, most important paper in science, as Watson's and Crick's famous paper on the structure of DNA was only two pages long). Because Satyendra offered no explanation of his thought process within the mathematical paper, it created a series of differing interpretations of his work, from being a "lucky guess," to having full intuition about this process. Ghose wrote that "Abraham Pais, a physicist and biographer of Einstein, noted that 'there had been no such successful shot in the dark since Planck introduced quantum in 1906.'" Other rumors suggest Bose didn't know the own importance of his new discovery (which clearly isn't true as shown in his correspondence with Einstein). People today still wonder how Satyendra Bose came up with his calculations, but that is a mystery we may never know. With the paper full of equations and having little explanation of his process, it would become an iconic work for both Satyendra and later, Einstein.

During this same time, Satyendra found out that his position at the University of Dhaka would not be renewed due to funding problems between the governments of Bengal and India. Simultaneously, Satyendra applied to study abroad in Europe for two years. His friend Saha had done this years before, after receiving his Ph.D., and was able to study in some of the most well-known laboratories in Europe. Because Satyendra was married, he was not offered the same opportunities. Due to these financial problems, some scholars believe that Satyendra desperately tried to use his new discovery to gain better opportunities. This shows that Satyendra did understand the importance of his own work for the international scientific community. He mailed his short paper to the *Philosophical Magazine* of London. When the paper was rejected by the magazine, Satyendra then sent the paper to Einstein.

Whether you believe that Satyendra did this as a last-ditch effort to save his academic career, or because he had seen Einstein's generosity before, one thing was clear: it was a gutsy move. Not only did Satyendra send Einstein the unsolicited paper, but he also tacked on a note, asking Einstein to translate it into German and publish it in the prominent journal *Zeitschrift für Physik*, if Einstein liked it. Bose additionally wrote: "I do not know whether you still remember that somebody from Calcutta asked your permission to translate your papers on Relativity into English. You acceded to the request. The book has since been published. I was the one who translated your paper on Generalized Relativity."

Gutsy move or not, Einstein was intrigued and did as Satyendra asked. Einstein translated and published Satyendra's paper, adding his own endorsement. A week later Einstein published his own paper that expanded on Satyendra's work. In the paper, Einstein used Satyendra's equations to predict the Bose-Einstein-Condensate or BEC (see chapter 10), a state of matter that, at ultracold temperatures, showed quantum interactions more predominantly. As Mukhopadhyay wrote in a 2021 article "In a normal gas, atoms move randomly at very high speeds, but in the case of the BEC, the atoms stay together in perfect harmony." After publishing Satyendra's paper, Einstein sent a short reply back to Satyendra in the form of a postcard. In it, Einstein gave Satyendra updates on the paper and invited him to Europe.

Einstein's postcard was apparently all it took. Once Satyendra showed it to the Indian authorities, he was given a traveling stipend and a generous payment for his family during the two years he would be away. He also didn't have to pay for the cost of his visa to Europe. A couple of colleagues wrote to physicist Paul Langevin in Paris to connect him with Satyendra. With these connections, his visa, and no doubt Einstein's postcard in his pocket, Satyendra boarded a steamer of the Lloyd Triestino line to Paris.

A European Tour

After some sightseeing, Satyendra finally landed in Paris in mid-October 1924. He lived in a boarding house that hosted many other Indian visitors, creating a familiar and comfortable community for Satyendra. In Paris, he quickly met Langevin, who introduced him to many famous physicists, including Maurice de Broglie (brother of Louis de Broglie) and Marie Curie. In fact, Satyendra would work in both Curie's laboratory as well in de Broglie's. From these positions, Satyendra became fascinated by the properties of crystals (a field known as X-ray crystallography) and saw how important it was to do experiments. He would put both interests to use when he returned to India. After about a year of staying in Paris, Satyendra wrote to Einstein that he would be in Germany soon and hoped to meet. With Einstein giving a positive response to the meeting, Satyendra set off for Berlin, nervous and excited.

Once in Berlin in 1926, Satyendra finally met Einstein. According to some sources, "When Einstein met Bose face-to-face, he asked him whether he had been aware that he had invented a new type of statistics, and he [Satyendra] casually said no." Einstein was curious about Satyendra and tasked him with helping to expand on new ideas that the physicist Werner Heisenberg had recently published. According to the 1974 Royal Society Memoir: "Bose obtained a letter of introduction from Einstein. With its help, he was granted certain privileges, such as the right to borrow books from the university [of Berlin] library and to attend the physics colloquium. 'I made good use of these privileges.'" As a person of color in Victorian Europe, Satyendra no doubt would have felt much discrimination from the people around him. Thanks to Einstein's letter, Satyendra was able to access more resources that would have been traditionally banned from minority groups like his own. Einstein also introduced Satyendra to many well-known physicists in Europe, including Lise Meitner, Fritz Haber, Otto Hahn, and others. Many of these scientists were rightfully impressed with Satyendra's abilities and enjoyed discussing problems with him.

Satyendra's visit to Europe illustrates what Dasgupta refers to as being the "peripheral scientist." Because science is both "an intellectual and social endeavor," scientists who are not a key part of the "central social circle" of learning are considered peripheral scientists. As Dasgupta explains, the two ways a peripheral scientist enters the main circle is by either solving critical problems facing that main circle or by being endorsed by someone within that main circle. Satyendra's case actually models both, as he expanded on Einstein's and Planck's work (without a Ph.D. or a physics background) and got approval from scientists like Einstein, Langevin, and others. Dasgupta believes that Satyendra was the ultimate scientist outsider as "Bose was a scientist who was entirely raised and trained in a colony, and thus a peripheral scientist in every sense of the term." No doubt Satyendra probably did feel somewhat like an outsider, as he saw very few other minority scientists during his visit.

Being a peripheral scientist probably also played a role in Satyendra's relationship with Einstein. It was not difficult to see that the relationship was rather one-sided, with Satyendra thinking more highly of Einstein than Einstein did of him. In all the letters Satyendra wrote to Einstein, he called Einstein his "master," reflecting the teaching structures of the East. "Bose's deep sense of gratitude to Einstein arose from the simple fact that, not only Einstein's theory, but Einstein himself, had made it possible for Bose to reach his goal of becoming a scientist," Dasgupta wrote. "Having projected onto Einstein a master-disciple relationship—a long-standing epistemic tradition from his own home culture—Bose perhaps considered himself bound by the 'loyalty' of that relationship, even when that 'loyalty' was going against his own interests." Because of his need for Einstein's approval, Satyendra took Einstein's rejection harder than normal. In 1924, after he wrote his first monumental paper, Satyendra sent two other papers to Einstein which worked on separate derivations by Einstein and other physicists. Einstein heavily critiqued these papers, and the "rejection" from his new "master" drove Satyendra to tear the third paper up entirely. We will never know what that third paper said, as no record of it remains. It would not be the last time Satyendra would destroy his work due to Einstein's influence. Many scholars put this down to perfectionism, as Satyendra couldn't bear his work not being perfect, especially to show to Einstein. Even if he wasn't a perfectionist, Satyendra's one-sided relationship with Einstein probably made things difficult.

As 1926 drove to a close, Satyendra began thinking about the logistics of returning to India. Surprisingly, Satyendra did not publish any papers while in Europe, though he worked with some of the world's leading physicists. Many experts use this as evidence for his "flash in the pan" contribution, and Einstein may have thought something similar, as some historians theorize that his eventual lack of interest in Satyendra and his work may have driven Satyendra to leave the field of quantum physics entirely. In truth, we will never understand the full dynamics between Einstein and Satyendra, but can only guess what happened based on what the research shows.

Before he left Europe, Satyendra had one more favor to ask Einstein. A professorship position was available at the University of Dhaka, and Satyendra's friends encouraged him to apply. However, because he didn't have a Ph.D., he was worried he would be rejected for the position. To better his chances of getting hired, Satyendra approached Einstein to ask for a letter of recommendation. Einstein was surprised, but quickly wrote a quick letter for Satyendra to send off:

> "The recent works of Mr. S.N. Bose, especially his theory of radiation equilibrium, signify in my opinion important and enduring progress of the physical theory. Also in personal discussion with Mr. Bose, I got the impression that he is a man of unusual gift and depth, from whom science has much to expect. He has also at his command an extensive knowledge and certain ability in our science. As a university teacher, he will certainly develop a successful prosperous activity."

Like before, this letter helped Satyendra obtain the position and by the end of 1927, he was the head of the Physics Department at the University of Dhaka.

Later Years and Legacies

Once at Dhaka, Satyendra helped build up the physics department from scratch, which included X-ray crystallography laboratories. "Bose set up a small chemistry lab and encouraged students to synthesize emetine, sulfa drugs, and other drugs," explained Ghose. "Much of this research was never published and the few papers that did appear hardly ever carried his name." Satyendra encouraged his students to build their own instruments with locally sourced materials, instead of being imported from foreign countries, reflecting the self-sufficiency he had witnessed during the earlier years of the Indian revolution. His colleagues recalled that Satyendra was working from 7 a.m. to late at night, with only a small lunch break. His office door was always open, and he made sure to answer any problems a student had, even if they were not in his field. Satyendra was noted as enjoying music and cats particularly and loved to discuss nearly any subject under the sun. He made sure all his students were heard and mentored. "He was also a dedicated teacher and mentor, and many of his students and coworkers were inspired by him over the course of his career," Mukhopadhyay explained. "For instance, Dr. Purnima Sinha was one of the first Bengali women to get a Ph.D. in physics. Her advisor was SN Bose. In the field of x-ray crystallography of clay minerals, she did a lot of important work. In 1990, she translated Erwin Schrodinger's *Mind and Matter* into Bengali. Her supervisor's legacy to spread basic science in Bengal inspired her to do this."

Similarly, the University of Dhaka was becoming more of a welcoming and supportive place. According to Wali: "The University had made great strides in providing higher education to Muslim students whose numbers increased from 160 in 1921 to 600 in 1941. The residence halls for Hindu students were originally run-on caste lines, with separate halls for Brahmins and non-Brahmins, were slowly being integrated." Unfortunately, these peaceful times would not last, as riots and famine broke out in the area. Worried, Satyendra moved himself and his family to the University of Calcutta in 1945.

The move to Calcutta would not be a peaceful one, as a period of intense violence quickly occurred. "Thousands of Hindus and Muslims were slaughtered in a three-day rampage in what has become known as 'the Great Calcutta Killing of August 1946,' an infamous tragic episode in India's history," Wali wrote. These killings included many students from the University of Calcutta, who were brutally murdered. No doubt this would have been a very difficult time for all those in the area, including Satyendra. The violence, unfortunately, continued, as India was declared fully independent on August 15, 1947, and Pakistan ceded from the country. This secession was particularly painful to Satyendra as part of his homeland of Bengal was now in a foreign country.

In 1956, Satyendra retired from the University of Calcutta. The university declared him a professor emeritus and he retained his office on campus, often continuing to work post-retirement. Though Satyendra had done decades of research, it would be only his first quantum physics paper that would be well-known to scientists outside of India. Satyendra even described his seemingly "one-hit wonder" career by later saying

> "On my return to India, I wrote some papers, I did something on statistics and then again on relativity theory, a sort of mixture, a medley. They were not so important; I was not really *in* science anymore. I was like a comet, a comet which came once and never returned again."

The other accolades Satyendra received over his life would disagree with his sentiment. He was awarded the Padma Vibhushan, India's second-highest civilian award, by the Government of India in 1954. He also became a Fellow of the Royal Society in 1958. He was also the President of India's National Institute of Science, as well as a member of the Indian Statistical Institute and the Indian Physical Society. He was also nominated for the Nobel prize on four different occasions but never received it. To Satyendra, these awards didn't really matter to him. He would be quoted saying "I have got all recognition I deserve." It was this humble attitude that made Satyendra stand out as a scientist. "The essence of Bose was to study and research science, without any expectations of attaining a celebrity status or a coveted recognition," explained Santanu Ganguly, an expert in quantum networking for Photonic Inc., in an interview with Kenna. "Humility and modesty led his meticulous way of working through intricate problems - a lesson for any researcher."

Throughout his later years as a professor, Satyendra emphasized the importance of teaching science in the native tongue of Bengali. He believed that this would help make the ideas more accessible and encourage more individuals to pursue science, which proved to be correct. Satyendra even published a magazine in Bengali that explained many of the ongoing projects at the University of Calcutta. Just before his retirement, Satyendra also worked on several papers on Unified Field Theory (see chapter 8). This theory combines the fundamental forces of physics with its elementary particles to create physical and virtual fields (obviously it's quite technical). During this time, Einstein was also interested in Unified Field Theory and Satyendra hoped to discuss his papers more with Einstein. He planned to meet Einstein at the 1955 International Crystallography Conference in Paris. He never got the chance, as Einstein passed away in April 1955. Hearing about Einstein's death, Satyendra was distraught and tore up the papers he was working on, again making them forever lost to time.

Satyendra himself passed away on February 4, 1974, at the age of 80. At his 80th birthday celebration, he remarked that he could "die now," as everything seemed complete to him. Those prophetic words haunted many for years after. In 1986, 12 years after his death, the Indian Parliament created the S.N. Bose National Center for Basic Sciences in Salt Lake, Calcutta. No doubt Satyendra would have been pleased with having a new place for students to learn.

While Satyendra is most closely tied to Einstein, English physicist Paul Dirac also played a significant role in establishing Satyendra's legacy. Dirac nominated Satyendra to the Royal Society in the 1950s, as well as coined the term "bosons" after Satyendra Bose. Dirac had the utmost respect for Satyendra. In the 1950s, when Dirac visited Kolkata to give a series of lectures, he spent a lot of time with Satyendra. In one story, Satyendra appeared to be sleeping in the front row during one of Dirac's lectures. "Writing an equation on the blackboard, Dirac seemed to hesitate and looked toward the white-haired Bose for continuation. Bose lumbered to his feet, scribbled the rest of the equation, and then resumed his somnolent posture," explained a 2007 *Physics Today* article. Dirac had many conversations with Satyendra about physics and came to admire Satyendra's brilliance even more during his visit. Mukhopadhyay told Kenna about how

> "Bose kept teaching at the University of Calcutta until 1956 when he was 62 years old and could finally stop. During this time, one of the founders of quantum mechanics, Paul Dirac, and his wife went to the Calcutta University building. They were riding with Bose in the same car. Bose let them sit in the backseat, which was nice and warm. He shared the front seat with the driver, which wasn't very big. But he asked for two more of his students to come in. Dirac was worried that the front seat might be too full. Bose turned around and said, 'We believe in the Bose-Einstein statistics,' which was his way of putting people at ease. Dirac then told his wife that in Bose-Einstein statistics, things are close to each other."

Thanks to the generosity and work of Einstein and Dirac, the story of Satyendra Bose has become more widespread. But within India, Satyendra is seen as a role model for many aspiring scientists. "Growing up in a Bengali family in Calcutta (present-day Kolkata), and having a physicist dad, I can't even remember when I first heard Bose's name, or even how many times I've heard his name, probably as many times as Tagore's" Abhijit Roy, a software engineer in Toronto, told Kenna. "Bengalis, being heavily influenced by art and culture, take a lot of pride in having been from the area that brought about the renaissance in India. There is also a lot of pride and love for the number of Nobel Laureates and Nobel nominees who are either directly from Bengal or have Bengali origins. I had books and a poster of Bose, along with some sports icons on my wall. Bose was someone you thought of when you tried to channel your mathematical mind. My dad would tell me the story of how Bose sent his paper to Einstein, after being rejected from a journal, and how Einstein understood what the papers held, and then the world got to know Bose. My father's eyes would light up every time he got to the end, no matter how many times he had talked about it. Bose inspired me to follow the sciences, learn about the universe, and learn about stars, atoms, and quantum theory. It's true that I'm not a genius like Bose was, but the role model that he is was part of the reason that I and many others have studied hard and been successful in life."

As Bose-Einstein statistics and the BEC are still being studied, Satyendra's work continues to live on and he still makes an impact on scientists around the world. According to Dr. Scott Aaronson, professor at the University of Texas at Austin: "Bose was the first person to clearly understand the statistics of the particles we now call bosons: the way swapping two identical bosons gives you exactly the same quantum state, which tends to decrease the number of possible configurations where the bosons end up in separate places, and thereby increase the probability that they're 'all piled up on top of each other.' In the 1920s, this work made a huge impression on Albert Einstein, who helped transform Bose from a young (relatively) unknown to someone whose name lives forever in one of the two basic types of particles of our universe." Bosons and the BEC continue to play a big role in quantum computing as well, as Aaronson explained that "there are some important quantum computing architectures and algorithms that do exploit boson and fermion statistics. And BECs have been explored as one possible building block for quantum computers, though it's not a leading proposal."

Things get complicated if we try to answer the question of *was Satyendra Bose just a "flash in the pan?"* While it's easy to find evidence that points one way or the other (and indeed Satyendra admitted he thought this about himself), it's important to put Satyendra's work into a historical perspective. Though he might have been in "the right place at the right time," in comparison to other scientists during this flourishing of physics, his work might seem to come up short. Because Satyendra lived during a time of a Renaissance for physics, especially quantum physics, the scientists he gets compared to seem almost mythical. Instead, if we were to separate Satyendra's career trajectory and research, we would find that it seems comparable to what a modern-day scientist does. Only a few scientists today may have such breakthroughs as Satyendra did, while most just research and publish papers throughout their lives without any big discoveries. If we understand this context, we can see that Satyendra was actually a very successful scientist whom history has interpreted in an unfair way. "The achievement and recognition that Bose received against all odds defines a legendary example of excelling," Ganguly added. "He did so, without any support from a system. A stellar example of rising against all odds by focusing on one's passion, aided by brilliance and work ethics." Perhaps with this retelling, we can rectify this story and help Satyendra be seen as the stand-alone genius that he truly was. While he may have been considered a "peripheral scientist" and an outsider during his time, Satyendra can become part of the "main circle" of key scientists through the retelling of his life and works, as his incredible story gets passed through the generations.

Photos & Images

(Right): A painting of al-Khwārizmī.

(Courtesy of the Australian Muslim Times)

(Left): Depiction of a library with students dating back to the 13th century. Experts believe this could be similar to what the House of Wisdom looked like.

(Courtesy of Zereshk and Wikimedia Commons)

(Right): some of al-Khwārizmī's mathematical notes.

(Courtesy of the Australian Muslim Times)

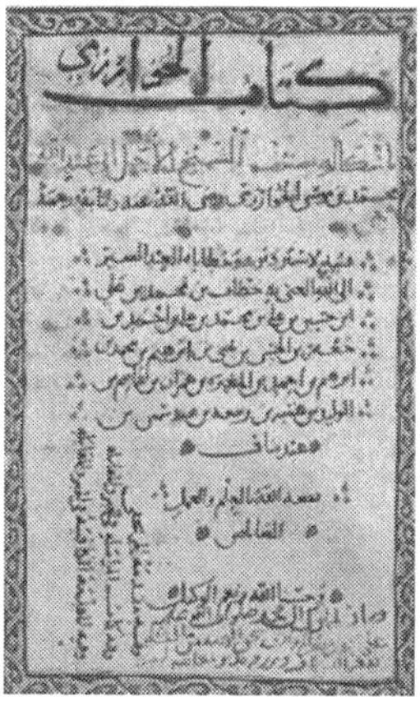

(Right): A painting of Gottfried Leibniz

(Courtesy of Wikimedia Commons)

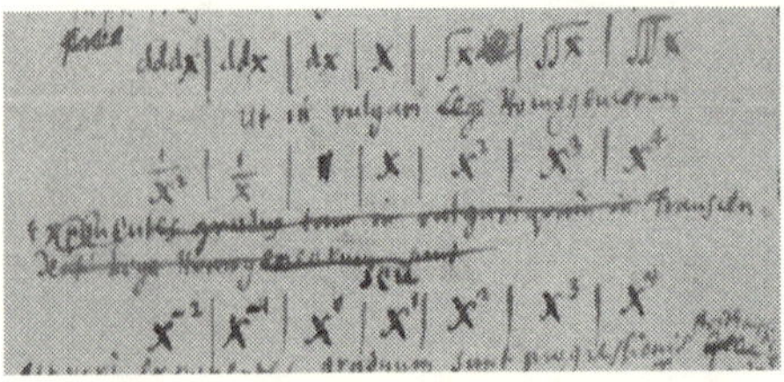

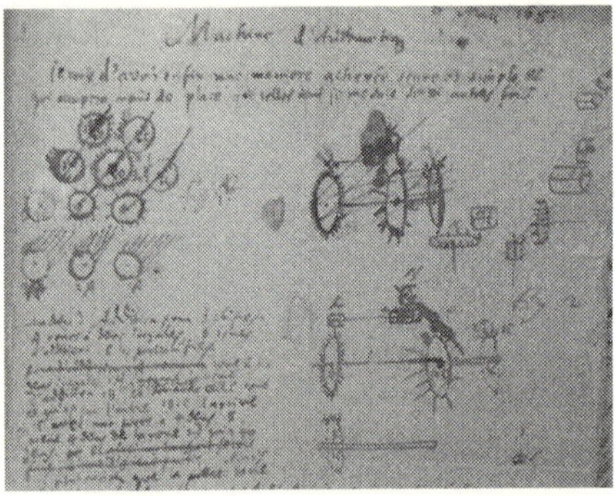

(Above left): Leibniz's calculus notations

(Left): Designs from Leibniz's journals for his calculating machine

(bottom left): A complete drawing of Leibniz's calculating machine

(Courtesy of: Stephen Wolfram, 2013, "Dropping In on Gottfried Leibniz", Stephen Wolfram Writings. https://writings.stephenwolfram.com/2013/05/dropping-in-on-gottfried-leibniz/)

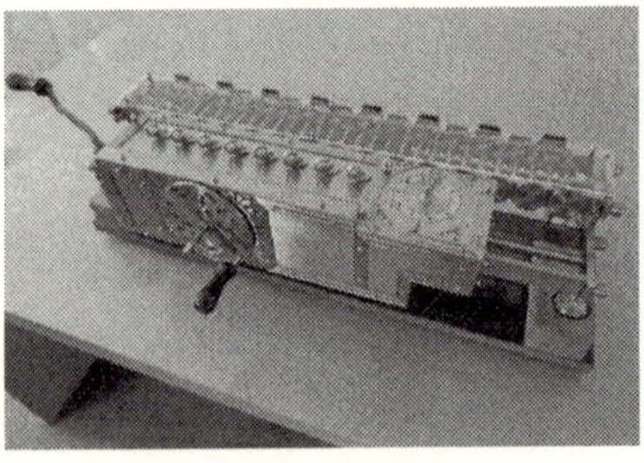

(Right): Leibniz's actual calculating machine at the Leibniz Museum in Leipzig

(Courtesy of Wuselig and Creative Commons)

(Left): A photo of Elmer Imes

(Courtesy of National Society of Black Physicists)

Virginia Love Pastor Gets Ten-Year Term in Pen

THE AFRO AMERICAN

Crowds Seek to Touch Becton's Bed

IMES MUST PAY

Rabbi Wise of N. Y. Speaks to 1,000

(Right): The 1933 edition of the *Baltimore Afro-American* publishes Imes' illicit affair scandal on the front page

(Courtesy of the Baltimore Afro-American archives)

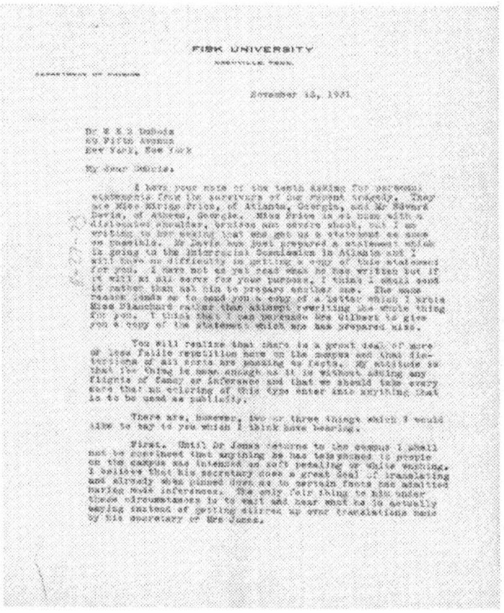
FISK UNIVERSITY

(Left): The 1931 letter Imes wrote to Harlem Renaissance writer W.E.B. Du Bois about the Julia Derricotte case

(Courtesy of the American Institute of Physics)

(Right): A photo of Satyendra Nath Bose

(Courtesy of Wikimedia Commons

(Left): Photos of Bose and Albert Einstein and their early letters to each other

(Courtesy of Kameshwar C. Wali)

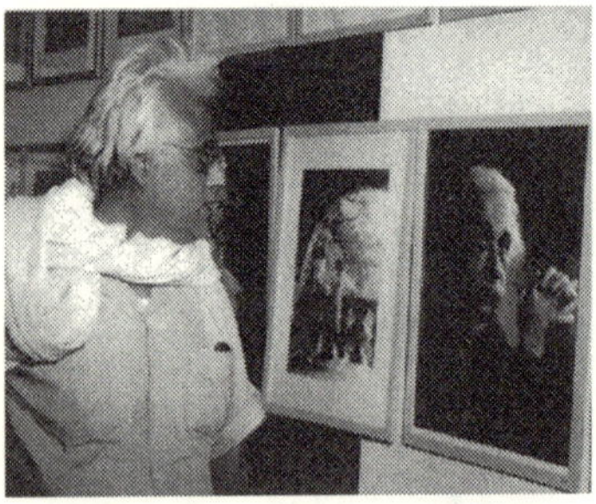

(Right): An older Satyendra Bose looking at a photo of Albert Einstein

(Courtesy of Kameshwar Wali and Etienne Eisenmann, found on OPTICA website)

(Left): Photo of Sin-Itiro Tomonaga

(Courtesy of Wikimedia Commons)

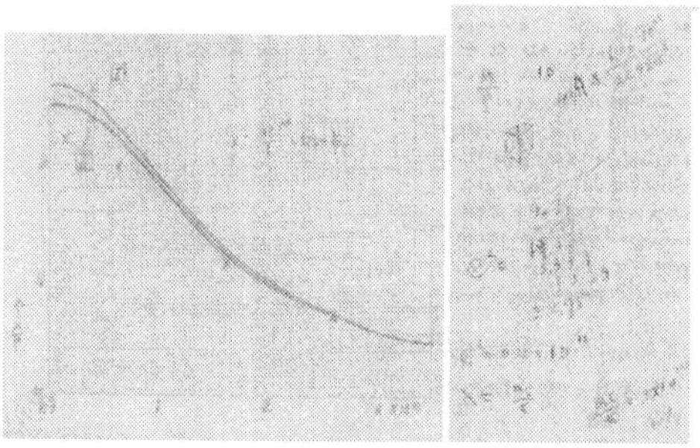

(Left): Part of the 1937 letter Tomonaga wrote to Hideki Yukawa about mesons

(Courtesy of the Yukawa Archive)

(Right): The three winners of the 1965 Nobel Prize in Physics

(Courtesy of Steve Hsu)

Sin-Itiro Tomonaga Julian Schwinger Richard P. Feynman

(Lower right): Tomonaga pours sake after winning the Nobel Prize

(Courtesy of Asahi Shimbun via Getty Images)

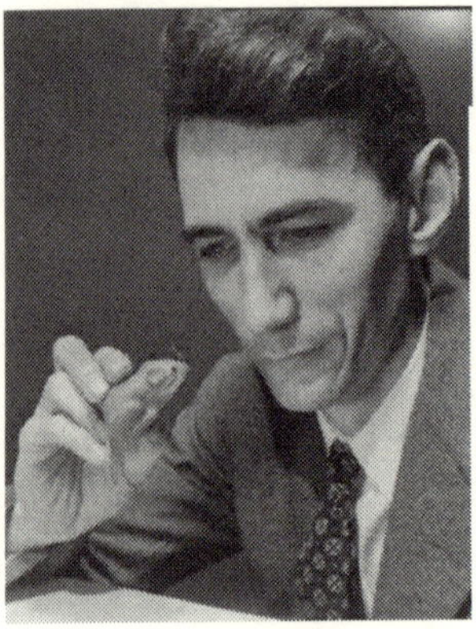

(Left): Photo of Claude Shannon playing with his mechanical mouse "Theseus" while working at Bell Telephone Laboratories

(Courtesy of Dobri Zheglov and Wikimedia Commons)

(Right): Shannon riding one of his famed unicycles while juggling

(Courtesy of the Shannon family and Jimmy Soni and Rob Goodman)

(Left): Shannon and his wife Mary Elizabeth "Betty" Moore Shannon building a chess-playing machine.

(Courtesy of the Shannon family and Jimmy Soni)

(Right): A diagram of Shannon's mathematical theory of communication

(Courtesy of Norm Friesen)

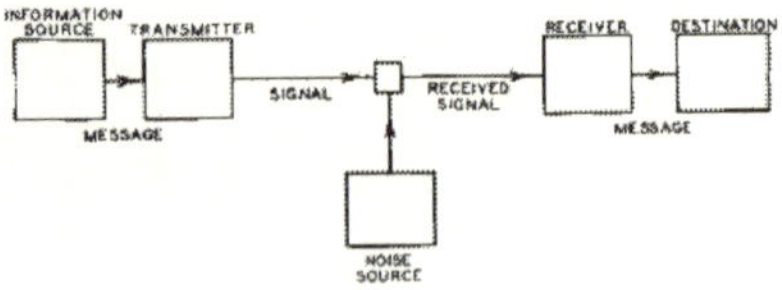

(Right): A photograph of Frances Elizabeth "Betty" Holberton (nee Snyder)

(Courtesy of the Holberton School and Priscilla Holberton)

(Left): A photograph of Monroe Snyder, Holberton's grandfather, who proposed setting up the NBS (National Bureau of Standards)

(Courtesy of William Snyder and Ancientfaces.com)

(Right): Part of the ENIAC machine at the University of Pennsylvania

(Courtesy of Penn Today)

(Left): Holberton (on the right) working on the ENIAC at the Moore School

(Courtesy of NIST)

(Right): Holberton working at the Remington-Rand corporation on the UNIVAC

(Courtesy of Hagley Digital Archives)

(Right): Silaban donning his usual uniform of long sleeve shirts and trousers with a cigarette

(Courtesy of Ensiklopedia Tokoh Batak)

(Left): Pantur Silaban teaching a lecture

(Courtesy of Tobatabo)

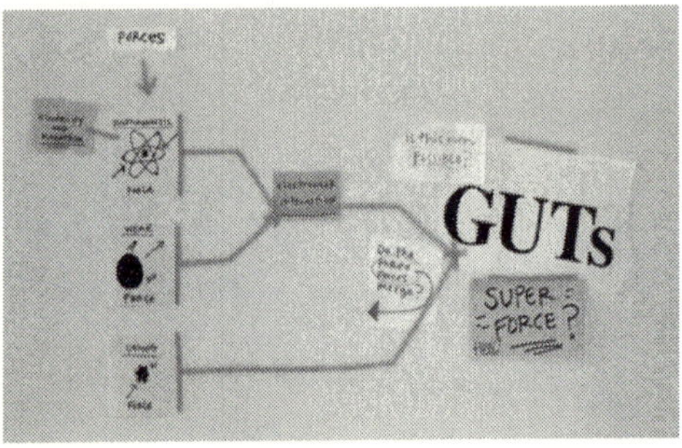

(Left): Showing how Grand Unified Theory (GUT) accounts for three of the four main forces of physics

(Courtesy of Symmetry Magazine and Sandbox Studios

(Right): A photo of Mark Reed with a nanodevice

(Courtesy of the Reed Lab at Yale University

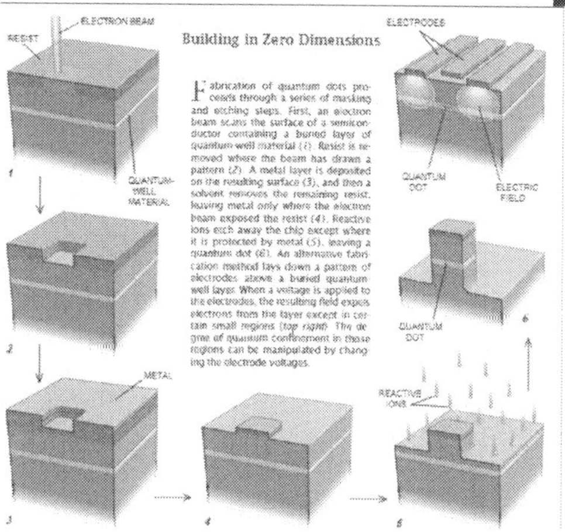

(Left): the process of creating a quantum dot via photolithography

(Courtesy of Mark Reed's Scientific American Paper)

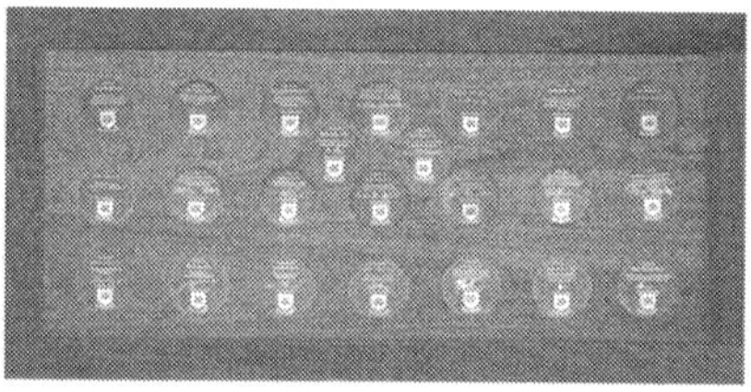

(Upper Left): The board that Reed made with a silicon wafer etched by photolithography for each of his students.

(Courtesy of Laurie Calvet)

(Lower Left) an etched board of the Reed Lab team, with Mark's signature cowboy look in the center

(Courtesy of Elizabeth Reed)

(Left): A photo of Deborah "Debbie" Jin in her laboratory

(Courtesy of NIST)

(Right): Jin with husband John Bohn (right) and colleague Brian DeMarco (middle) and baby Jaclyn.

(Courtesy of Brian DeMarco)

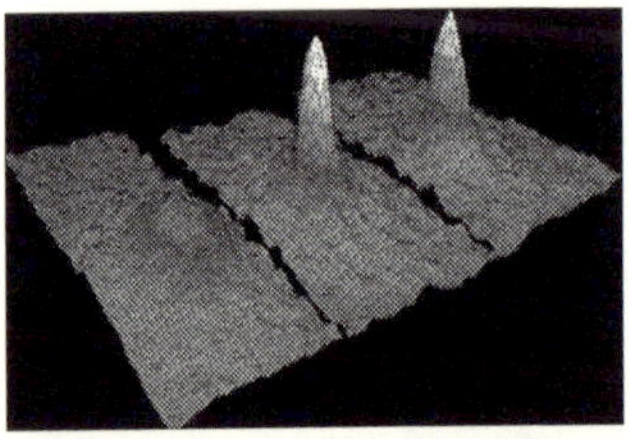

(Left): Images of the Bose-Einstein Condensate at various temperatures

(Courtesy of NIST, JILA)

(Left below): Jin and Ye awarded as 2018 "Highly Cited Researchers"

(Courtesy of JILA)

(Bottom Left): The 2007 Championship JILA Monster team. Front row, left to right: Carrie Weidner, Laura Sinclair, Celeste Landry, Jackie Bohn (mascot), Debbie Jin, Heather Lewandowski, Karen Bryant. Back row, Eric Cornell, Don Woodraska, Jayson Stewart, Juan Pino, John Bohn, Russ Stutz, Aaron Leanhardt.

(Courtesy of John Bohn)

SECTION 2 – Post-World War II Giants

Chapter 5

Sin-Itiro Tomonaga - "The Thinker"

One of the students coughs in the grimy classroom. Or at least, it's supposed to be a classroom. The dilapidated building seems as though it could fall onto the small circle of students at any moment. While they may be worried, their teacher, physicist Sin-Itiro Tomonaga, tries not to notice as he writes on a cracked piece of a small slate blackboard. The chalk squeaks at a level that seems almost too loud in their quiet surroundings, Suddenly, an air raid siren erupts nearby, its shriek causing the students to jump. Sin-Itiro pauses his writing and then puts the blackboard down. He silently motions his students to follow him, as they make their way into a connecting tunnel. Sin-Itiro's family is huddled to one side and the students huddle against the adjacent wall, all looking up, waiting. Only Sin-Itiro remains standing, trying to keep calm. He hopes that this is just another drill and that the bombs will stay far, far away.

***

As you've probably noticed throughout this book, and will continue to notice, war and science are often linked. And with Sin-Itiro Tomonaga's story, they become even more so. Stories from Japan during World War II are not as well-known as stories from the West, which can make it difficult to see the full picture and just how catastrophic the War was to Japanese culture. Looking at the life of Sin-Itiro Tomonaga, we can finally begin to understand how World War II affected scientific research and the science community as a whole. This became especially important for quantum physicists like Albert Einstein, Richard Feynman, J. Robert Oppenheimer, and others as well. We picked Sin-Itiro as one of our giants for multiple reasons, the biggest being that his story rarely gets told (in fact this book is one of the first to analyze his work). He not only won the 1965 Nobel Prize in Physics along with Julian Schwinger and Richard Feynman but also dedicated the latter part of his life to fighting against nuclear weapons while promoting safe research into nuclear energy. Sin-Itiro's method for research was also somewhat unorthodox, as he had his students play a valuable role in discerning new theories. His thoughtful demeanor and generosity endeared him to his students and colleagues and made him an inspiration for many generations of physicists.

As there are few sources on Sin-Itiro (and even fewer in English), and many contradict each other or suggest different dates for differing events, we will do our best to piece together the story of this brilliant scientist and caring individual. Like Satyendra Bose, Sin-Itiro was a theoretical physicist, so we have no laboratory experiments to show for his work. Instead, we have his theories, which are so incredibly technical that many would avoid explaining them altogether. In telling Sin-Itiro's story, we will do our best to take on this challenge and illustrate his work as best we can (though neither of us has a background in physics). As you'll see from our writing, Sin-Itiro's work did help in part with advancing the possibilities of quantum computing.

Born on March 31, 1906, in Tokyo, Sin-Itiro would grow up (like most of our giants) in a family that enjoyed learning. His parents were Sanjuro and Hide Tomonaga. Hide, his mother, was the daughter of a feudal clan leader, but also was somewhat educated, as she was "a graduate of Ochanomizu Girls Middle School," according to the book *Science and the Building of Japan.* Sanjuro, Sin-Itiro's father, was a professor of philosophy, first at a university in Tokyo, then at Kyoto Imperial University. According to a *Physics Today* book review article concerning Sin-Itiro: "His father, a close friend of the chief priest of the head temple of the Tendai Buddhist sect, was invited to live in a large house on the temple grounds. The house became filled with books, which nurtured the young Sin-Itiro's passionate curiosity, and the temple grounds helped nourish his intense communion with both nature and Japanese culture."

When his father moved to Kyoto, he left Sin-Itiro behind with Hide, only to be reunited again in Kyoto in 1913. This must have been very hard on Hide, as Sin-Itiro was a weak and sickly boy. Even his school teachers noticed his physiology, as his kindergarten teacher later met Sin-Itiro, decades after his schooling and remembered him for it.

> "Even though the exchange took place some 60 years after Tomonaga attended the school, and even though the teacher had taught over a thousand children in her 40 years as a kindergarten teacher, she remembered him very distinctly as a thin, bony, extremely quiet little boy who was always on the verge of tears," the *Physics Today* article stated.

The lack of a present father plus a sickly disposition could have created many personal and mental issues for Sin-Itiro, especially since he was also bullied at school, but surprisingly they didn't. He was described as confident, even if a little too much. Sin-Itiro was fascinated by science and originally wanted to be a biologist. The book *QED and the Men Who Made It* recalls various experiments that Sin-Itiro did as a young child. From filling balloons with hydrogen gas obtained by splitting water molecules to building an electric doorbell from scraps with a friend, Sin-Itiro found himself always busy tinkering with things to figure out how they worked. "Once, out of curiosity to see what would happen, they plugged the electric doorbell directly into the wall outlet: the resulting meltdown remained a well-kept secret between them," stated *QED*. He even created photographic film slides using gels he distilled from seaweed. Sin-Itiro's curiosity would be further piqued in 1922 when Albert Einstein visited Japan.

By the time Einstein arrived in Japan, he was already a well-known scientist due to his theory of general relativity. On October 8, 1922, Einstein and Elsa (his second wife) sailed aboard the *SS Kitano Maru* from Marseille to Japan. The trip took nearly six months as they visited exotic locations like Egypt, Hong Kong, and Singapore before arriving in Japan. About the trip, Einstein wrote in his diary that "The invitation to

Tokyo pleased me a great deal, as I have been interested in the people and culture of East Asia for a long time." Einstein drew thousands around Japan to him as he traveled, As *Smithsonian Magazine* stated: "In Japan, thousands packed auditoriums to hear him expound on his theory of relativity for three or four hours at a stretch, in remarks laboriously translated from German." One of those interested fans was Sin-Itiro, who read every press article about the visit.

Sin-Itiro was not pleased, however, with how the press explained Einstein's theories. "Being a 'cocky middle school student,' he turned to a book by Jun Ishihara and became fascinated by the four-dimensional world and by non-Euclidean geometry," explained the *Physics Today* article. "He decided to study physics." To understand the field further, Hide used her networking skills to connect Sin-Itiro with physicists in Tokyo. Hide's passion for her son's education no doubt helped tremendously in propelling him forward into his career. Sin-Itiro would later say: "My career in physics came about because of Einstein's visit to Japan in 1922, when I realized that the world of physics holds many wonders. I was enchanted by the possibility of investigating this fascinating world."

Sin-Itiro went to Kyoto Imperial University for his undergraduate degree in physics, the same university his father taught at. We don't know if Sin-Itiro interacted with his father at all, or what sort of relationship they had at this time. We do know that Sin-Itiro could not have picked a better time to study physics. "At the time physics was being revolutionized by such scientists as Paul Dirac, Werner Heisenberg, and Wolfgang Pauli, who were forming the backbone of quantum mechanics," stated a 2012 Magnetic School article. These three theoretical physicists would each play key roles in Sin-Itiro's future work on quantum mechanics. Dirac would formulate a wave equation, which helped connect relativity theory with quantum theory, something Sin-Itiro would expand on later. Heisenberg found the "Heisenberg Uncertainty Principle" (see chapter 9) and would work directly with Sin-Itiro for a time. Pauli established many principles and rules, but for Sin-Itiro's work specifically, Pauli's most relevant theories connected the spin of a particle with "quantum statistics," such as a particle's internal state. Of course, all three men had Nobel Prizes in Physics, which gave Sin-Itiro hope he too may be awarded one someday.

After graduating with his bachelor's in physics in 1929, Sin-Itiro remained at Kyoto Imperial University to get his Ph.D., as Japan was in the middle of an economic crisis and no jobs were guaranteed. It was during this time that he heard a lecture by professor Yoshio Nishina at The Institute of Physical and Chemical Research, or the Science Research Institute (better known by its Japanese acronym, RIKEN). Nishina was already a well-established quantum physicist, having studied under British physicist Ernest Rutherford in the U.K. and also working with Niels Bohr in Copenhagen. After returning to Japan in 1928, Nishina convinced Heisenberg to come and give several talks at RIKEN. In 1929, Dirac also gave a talk at RIKEN, which Sin-Itiro watched. To say that Nishina was well-connected was an understatement, and for Sin-Itiro, Nishina would be the perfect mentor. Nishina was likewise fascinated with Sin-Itiro, seeing his potential to be an influential scientist in the future. Sin-Itiro joined Nishina's research group at RIKEN as a paid assistant.

There, he and Nishina would work on expanding previous quantum theories, as Sin-Itiro became the "house theorist," according to *QED*. Nishina would also become a father figure for Sin-Itiro, offering him life advice as well as career guidance. Nishina's laboratory was a hub of activity in the 1930s. It was where the first investigations into nuclear physics within Japan took place. The laboratory created a close-knit community. As Sin-Itiro later recalled in 1968: "We all got together after lunch every day, an eager group of people discussing various matters, not only physics but also such things as plans for beer parties, excursions, and so on..." At RIKEN, Sin-Itiro also learned how to "get the numbers out," or results from his work. As a theorist, Sin-Itiro would be researching previous physics theories and expanding mathematical calculations to try to simplify these theories, or connect them to other theories, or try to predict other quantum behavior.

The relationship between theorists and experimentalists is rather nuanced, as each plays a range of roles in advancing scientific knowledge. A theorist may try to expand on previous theories or use experimental results to improve a theory's predictability in order to mathematically explain what's happening. Similarly, experimentalists can use physical theories in many ways, such as a type of laboratory protocol, to explain the results of an experiment. As both of these types of scientists work to advance our understanding of the world, they often collaborate on different papers or projects.

Dirac would be one such theorist that Sin-Itiro would later pull from for his own research, In fact, Sin-Itiro, Nishina, and other RIKEN researchers loved Dirac's theories so much that they spent a significant portion of 1935 translating Dirac's work into Japanese, making it more accessible for other researchers.

In 1937, as part of an agreement between RIKEN and the University of Leipzig, Sin-Itiro traveled to Germany to study with Werner Heisenberg. There, Heisenberg encouraged him to expand on some rather confusing theories. After a year or so of work, Sin-Itiro felt depressed and was worried his research was stuck. This actually happens more often than expected as science is a field of trial and error. Sin-Itiro wrote in his diary at the time: "As I went on with the calculation, I found the integral diverged-was infinite. After lunch, I went for a walk. The air was astringently cold and the pond in Johanna park was half-frozen, with ducks swimming where there was no ice." It is rather common for young researchers, particularly graduate students, to feel depressed. According to a 2021 Harvard study, "graduate students are six times more likely to experience depression and anxiety than the general population, with a whopping 36% of them have sought help for depression and anxiety." Sin-Itiro certainly felt this depression while trying to achieve his Ph.D. In his diary, he added that: "Returning home, I read a book on physics. I don't understand it very well. Meanwhile, I suffer...why isn't Nature clearer and more directly comprehensible?"

Thankfully, Sin-Itiro had mentors who encouraged him to continue forward in his research. Nishina wrote Sin-Itiro a letter in which part of it read:

> "We stand at the crossroads, not knowing what lies before us. Whether we go left or right depends upon luck and our feelings at the time. Even if there is a big

> difference if one proceeds along a certain path, there's not too much worrying about that. In time, one will get lucky and things will go right."

Heisenberg, for his part, echoed a similar sentiment to Sin-Itiro. He explained that "It is very common for a physicist to start from a very crude model, without knowing whether his assumptions are valid. You have no need to worry about your result." Heisenberg also encouraged Sin-Itiro to publish his work, raising the student's spirits.

In 1937, both Sin-Itiro and Heisenberg began looking at meson theory. This predictive theory was established in 1935 by Hideki Yukawa (who ironically enough went to high school with Sin-Itiro and would also be awarded a Nobel Prize). According to Yukawa's theory, nucleons (made of protons and neutrons within an atom's nucleus) interact with a strong force, known as the nuclear force, via subatomic particles known as mesons. Sin-Itiro studied Yukawa's theory, and in a long letter to Yukawa looked at the mathematical structures of the theory, like the reactions coming from the interactions between mesons and nucleons, further clarifying the theory. The actual existence of mesons wouldn't be experimentally confirmed until 1947 by physicist Cecil Frank Powell, but this initial theory allowed Sin-Itiro an opportunity to begin his own projects. His work on this theory would be a core part of his Ph.D. thesis, which he would successfully complete in 1939.

In 1939, Sin-Itiro was forced to return to Japan due to the start of World War II. No doubt he would have witnessed the tell-tale signs that war was on the horizon. Germany by this time seemed to be completely under Hitler's spell, and many in the scientific community worried about what war would do to their research. Fleeing on a ship bound for Japan, Sin-Itiro no doubt wondered the same thing. World War II would completely isolate the East from the West, forcing these two communities of scientists to work independently of each other. For Sin-Itiro, this meant that he would never see Heisenberg again.

Years of War

After returning to Japan in 1939, Sin-Itiro received his Ph.D. and in 1940, became a professor at the Tokyo College of Science and Literature. His students absolutely loved him, as he would make sure to walk them through difficult physics problems. One previous student recalled a 1939 lecture Sin-Itiro gave at RIKEN, where he spoke for hours, writing advanced equations, without relying on any notes. It was just one example of how Sin-Itiro used his genius to help others learn. He would often say: "If you formulate the problem correctly, that is, if you ask the right question, the answer emerges spontaneously."

1940 was also the year Sin-Itiro would marry Ryoko Sekiguchi. According to the book *Science and the Building of Japan*: "His new wife was the daughter of R. Sekiguchi, who had been director of the Tokyo Metropolitan Observatory." The couple would have two sons and one daughter together. Sin-Itiro often called Ryoko his "guardian," and the couple remained happily married until Sin-Itiro's death in 1979.

The early 1940s became a time of great importance for Sin-Itiro and his students as they would formulate the super-many-time theory. This theory was an advancement on Dirac's many-time theory published in 1932. "In this many-time theory developed by Dirac, electrons were treated according to the particle picture," Sin-Itiro explained in his 1965 Nobel Prize lecture. "Alternatively, in quantum theory, any particle should be treated according to the wave picture. As a matter of fact, electrons were also treated as waves in the Heisenberg-Pauli theory, and it was well-known that this wave treatment was frequently more convenient than particle treatment. So, the question arose as to whether one could reformulate the Heisenberg-Pauli theory in a way which would be more satisfactory relativistically when electrons were treated as waves as well as the electromagnetic field." In looking at Dirac's many-time theory, which helped connect two previous theories together, Sin-Itiro focused on the time aspect. "Tomonaga was guided by his conservative stance and believed that advances were made by limited, incremental, evolutionary steps," stated *QED*. Dirac's theory, which tried to predict quantum dynamics, suggested a "same instant of time" for multiple factors, which led to problems with calculations. Instead, Sin-Itiro expanded on this idea by suggesting that each point in space had its own specific time, simplifying the theory. He was able to publish his theory, now called the super-many-time theory in 1943.

By 1943, the war was hard to ignore. Many scientists had been asked by the Japanese military to abandon their research to help research new weapons. Sin-Itiro found that many of the students at the University were being drafted into the Japanese military. The students who weren't drafted often suffered an entirely different fate, as their homes were destroyed during air raids by the Allies. The war also created food shortages and rationing. Between a gnawing hunger and a half-burnt city, many in Tokyo were depressed and distraught. Sin-Itiro did his best to help lift his students' spirits by offering special lectures (either on Fridays or Sundays depending on your source) to discuss quantum physics. "Hayakawa Satio, who was a student in the course, remembers that during the last lecture, Tomonaga wrote down the names of some prominent physicists—Bohr among them—and their age when they made their important contributions," explained *QED*. "All of them were young, and Tomonaga encouraged his audience to follow in their footsteps. He then wrote down Planck's name and that of some other physicists who had made their important contributions when they were older, and he encouraged himself to emulate them." For many of his 30 or so students, this helped immensely in easing their worries.

While the end of World War II brought celebration for many, it created confusion and chaos in Japan. Both Hiroshima and Nagasaki were completely destroyed. Food shortages were even worse, and many sought shelter where they could. "It was in the darkest period after the war that young physicists met together in the Tomonaga Laboratory," explained a 1966 *Progress of Theoretical Physics* article. "A few burnt buildings in the midst of ruins which originally belonged to the Research Institute of the Army. The place where Tomonaga lived was called 'Tomonaga house' which, however, hardly deserved the name of 'house' since it consisted of one room built in a cave opening into a tunnel originally built for rifle tests." In these dingy quarters, Sin-Itiro would continue teaching students and working. Worn out by the war, he decided to translate all his works into English, as it required "little thinking." According to *QED*, "he himself typed the translations on the backside of previously used papers." To help solve the rampant food shortage, Sin-Itiro investigated photosynthesis at RIKEN to find ways to grow more food. In 1946, he returned to his work in quantum physics. There, he worked with his students to use the super-many-time theory to expand on other aspects of quantum physics. It would only be in 1947 that he would do the important work that would later earn him a Nobel Prize.

Renormalization Theory and Visiting America

During the late 1940s, physicists were beginning to realize there were big problems with some of the earlier proposed theories. One of these theories was the hydrogen-Schrödinger equation, which stated that the energy levels of the hydrogen election depended on a factor called a "principal quantum number" (which is rather difficult to explain). Unfortunately, this was not accurate to what had been observed. As physicist Willis Lamb discovered in 1951, there was a shift in energy levels, called the Lamb shift. Even before Lamb published his results, physicists had already begun to see that hydrogen was unique, and created holes in previous theories. Sin-Itiro and his group decided to try and mathematically solve these issues. "I mobilized young people around me and we performed the computation together," Sin-Itiro stated at his Nobel Prize lecture in 1966.

In looking at the theory, Sin-Itiro realized that many of the mathematical values were infinite, which would make it harder to solve. To cancel out the infinites, he and his students used separate experimental values for the mass of the electron and charge within the equation. As Sin-Itiro added in his Nobel Lecture: "When a theory is incompetent in part, it is a common procedure to rely on experiment for that part. The procedure is called renormalization of mass and charge and our method has brought the possibility that the theory will lead to finite results by renormalization, even if it contains defects." Using the experimental substitutes of the mass and charge of hydrogen's electron, the infinites cancel out and the theory becomes more explanatory. As writer Charlie Wood explained in his article for *Quanta Magazine*: "Mathematically cutting the sums short and shuffling the infinity around, physicists now understand, is the right way to do a calculation when your theory has a built-in minimum grid size." Ignoring the infinites is not only a process used in science but also in many other industries. "When engineers design a skyscraper, they ignore individual molecules in the steel," wrote Wood. "Chemists analyze molecular bonds, but remain blissfully ignorant of [subatomic particles] quarks and gluons." In implementing renormalization into the theory, Sin-Itiro was able to make the equation clearer and more straightforward. Feynman later referred to the entire process as "hocus-pocus," but did admit it helped get the much-needed results. As Japan had been isolated from the West, news on the Lamb shift had not arrived until much later. By then, Sin-Itiro was able to use Lamb's data to improve his theory even further. Other physicists like Julian Schwinger and Richard Feynman were also working on this problem independently, and all three would be awarded the 1965 Nobel Prize in Physics for their work in quantum electrodynamics (QED).

QED plays an important role, not just in quantum computing but in everyday life. As associate professor Yidong Chong of the Nanyang Technical University told Kenna: "QED is the theoretical physics framework for describing how electrons (or other charged particles) interact with the electromagnetic field at the fundamental level. It underlies our theoretical understanding of basically every kind of everyday physics. The emission and absorption of light, why you don't fall through the floor, why atoms form chemicals and undergo chemical reactions, all these are governed by electromagnetic laws which are ultimately contained in QED."

QED is also a key component for many technologies, including quantum computing. "QED is important for quantum technologies simply because electrons and electromagnetic fields are our basic ingredient for interacting with the natural world," Chong added. "So, if you are using a quantum technology of any sort, you are making use of some formulation of QED. But, on some quantum technology platforms, the connection is more direct because they make use of a form of QED called 'circuit QED' which describes the quantum mechanics of electrons moving in electrical circuits." These electrical circuits are essential to advancing quantum computers. According to Lorraine Tsitsi Majiri, president of the quantum computing organization OneQuantum Africa: "Quantum electronic circuits are a key component of quantum computers, which have the potential to revolutionize the field of computing."

One of the major areas of progress in QED in 2022 is in the field of precision measurements. QED calculations have been used to make incredibly precise measurements of the properties of subatomic particles, such as the electron's magnetic moment and the fine structure constant (both inherent properties of the electron). These measurements have been used to test the predictions of QED and to search for new physics and in the field of quantum gravity, QED calculations are being used to test theories of quantum gravity and to search for new physics beyond the standard model.

Sin-Itiro had no idea how important his work in QED would become when he began working on it. Instead, he wrote to well-known physicist J. Robert Oppenheimer about his research, results, and challenges. Oppenheimer was previously the director of Los Alamos National Laboratory (see chapter 7) and in 1948 was the head of the Institute for Advanced Study at Princeton University (see chapter 6). His work of course was significant in the making of the atomic bomb, which caused many to either view him as either a savior or a murderer. After reading Sin-Itiro's letter, Oppenheimer wrote a reply suggesting that Sin-Itiro write a summary of his research and publish it in the journal *Physical Review.* He also invited Sin-Itiro to the Institute for Advanced Study as a visiting scientist. Excited, Sin-Itiro followed Oppenheimer's advice, and before he knew it he sat in a third-class cabin aboard a steamer ship headed for America.

We don't have many records of what Sin-Itiro actually worked on at the Institute, but we do know that the Institute was a hub of scientific activity during the time. It had been founded back in 1932 on two principles: "those individuals who work at the Institute should be selected on the basis of their abilities alone and with no regard to race, religion, or gender; and the Institute should enable the curiosity-driven pursuit of knowledge with no view to its immediate utility or the expectation of meeting predetermined goals." With these two principles, which were quite progressive for 1930s America, the Institute welcomed scientists from all over the world. The previous faculty included Oppenheimer, but also Einstein, von Neumann, and others. Sin-Itiro no doubt thrived during his visit there. He did give several guest lectures which were well attended. Because Japan and the rest of the East were finally open after World War II, Sin-Itiro's visit illustrated the bonds reforming between the two spheres, especially for the scientific community.

A Journey into Science Policy

In 1950, Sin-Itiro returned to his beloved Japan. According to *Science and the Building of a New Japan*: "He returned to Japan a new man, wearing fashionable suits made in the United States and smiling with a new set of false teeth." He was ready to get back to his research, but having been somewhat changed by American culture, Sin-Itiro also decided to shift his research focus. "The Kyoto spirit dictates that if everybody is doing the same thing, you do something else," explained *QED*. "By the fall of 1948, it seemed to Tomonaga that everybody was doing quantum electrodynamics and that it was time therefore to leave this fashionable field." Shifting his focus, Sin- Itiro decided to turn to a growing concern for many scientists around the world, atomic energy.

With the introduction of the atomic bomb in 1945, the world was ushered into an atomic age. Many countries began to push for more research on atomic energy, hoping to create their own arsenal of bombs as well as possible sources of clean energy. For many scientists, this was concerning, as some initial steps were being skipped in the stages of nuclear testing. Sin-Itiro famously stated during this time: "The promotion of research into atomic energy is like trying to build the second story before the foundations are firm." He decided that it was time to establish proper regulations for nuclear testing. To do this he needed to transition into science policy.

Science policy focuses on the creation and maintenance of regulations for scientific research. Science policies can target a wide range of issues, from the ethics of animal testing to the use of pesticides. Because science policy is often handled by politicians or other government figures, few scientists want to be a part of the policy-making process. However, many experts have urged for more scientists to be involved, as they have expert knowledge that is crucial to developing and maintaining policies in a sustainable way. For Sin-Itiro, the transition from research to policy may not have been as straightforward. It became even harder after Nishina suddenly passed away in 1951. Sin-Itiro was shocked and completely devastated by the loss of his father figure and mentor. Because Nishina was heavily involved in many science policy committees, Sin-Itiro took over Nishina's role for many of these positions, learning quickly how the policy process worked.

This was just in time, as Japan was undergoing a huge atomic phase. 1954 was an especially atomic year as "In March, the Japanese government went boldly ahead with its first budget allocation for atomic energy without proper consultation with physicists or the general public," explained *Science and the Building of Japan*. "At the same time, the United States began H-bomb tests in the Pacific." Not surprisingly, scientists around the world were horrified by these global atomic tests. They felt responsible for bringing about the atomic age and wanted to do something about it. The atomic bomb also changed how the public saw scientists. Instead of being seen as trustworthy figures, much of the general public viewed scientists as cold and cruel. This trend was picked up by popular science fiction during the 1950s and onward, as well as the increased use of atomic bombs as plot devices. For many scientists, something had to be done to restore trust and make sure nuclear research was conducted in a safe manner.

In 1955, Albert Einstein and Bertrand Russell created the Russell-Einstein Manifesto, which was published on July 9, 1955. It would be the last public document Einstein signed before his death, and it was a warning to all of humanity about the development and use of nuclear weapons. Part of the Manifesto read: "In the tragic situation which confronts humanity, we feel that scientists should assemble in conference to appraise the perils that have arisen as a result of the development of weapons of mass destruction, and to discuss a resolution in the spirit of the appended draft." Following this Manifesto, in 1957, Russell invited scientists from all around the world to Pugwash, Canada, for a meeting. This meeting would later be called the Pugwash Conference. Sin-Itiro attended the first conference, along with a couple of other prominent scientists to try and figure out what could be done about nuclear testing. Another part of the Manifesto reflected the serious atmosphere of the Conference, as it recalls: "We appeal, as human beings to human beings: Remember your humanity and forget the rest. If you can do so, the way lies open to a new paradise; if you cannot, there lies before you the risk of universal death."

For Sin-Itiro and the other Japanese physicists, they were rightfully worried about the consequences of atomic energy testing in Japan. Writer Makinosuke Matsui, in *Sin-Itiro Tomonaga, the Life of a Japanese Physicist*, describes the general concern for Japan's environment: "In those days, as a consequence of the nuclear detonations that had been conducted one after another in the atmosphere, radioactive fallout continued to rain down over Japan, and the extent to which it was harmful was a matter of serious concern." To fully investigate, Japan pulled out of the conference and formed its own group, with Sin-Itiro as Chairman. He worked tirelessly for the rest of his life to research the current destruction by nuclear tests, but also to advise Japan's government to create important regulations around nuclear testing. Thanks to his work he was able to help establish the Institute for Nuclear Study at the University of Tokyo.

Nobel Prizes and Later Years

In 1965, Sin-Itiro received life-changing news. He had been awarded the Nobel Prize in physics. In his 1966 Nobel lecture, he downplayed his role in the research, stating "So far, I have told you the story of how I played a tiny, partial role in the recent development of quantum electrodynamics," While Sin-Itiro did in fact play a larger role, he also worked with his team of students who helped advance the science with him. Additionally, fellow Nobel laureates Julian Schwinger and Richard Feynman also played their own roles in advancing QED.

Besides Einstein, Feynman is probably the most well-known quantum physicist to date. With clear communication skills and a charming personality, Feynman helped to make this technical field more understandable, inspiring generations of future scientists. Feynman helped work on the Manhattan project during World War II, and after that taught, first at Cornell University, and then at the California Institute of Technology (CalTech) as a professor. Between 1961-1963, Feynman gave several lectures at CalTech to college freshmen and sophomores. These lectures became compiled and published into a book: *The Feynman Lectures of Physics.* This book is still considered to be the most popular physics book ever published. Besides winning the Nobel Prize, Feynman became well-known for his help in investigating the Challenger space shuttle explosion in 1986. He, unfortunately, passed away in 1988 from abdominal cancer. To share a Nobel Prize with someone as influential as Feynman no doubt made Sin-Itiro feel incredibly honored.

The Nobel Prize would not be the only award Sin-Itiro would receive for his work. He was also awarded the Asahi Press Award (1934), the Japan Academy Prize (1948), the Lomonosov Medal from the Academy of the U.S.S.R. (1964), and the Grand Cordon of the Order of the Sacred Treasure (1976). These awards were given for Sin-Itiro's many different research projects, of which we've only investigated a few in this chapter, as there are too many to explain in just a handful of pages. With these awards and his role in science policy, it was no surprise that Sin-Itiro was chosen for many leadership roles.
He served on many influential governmental committees, eventually becoming the President of the Science Council, and he was frequently appointed the official emissary to represent Japanese culture and science at international gatherings, explained the *Physics Today* article. Sin-Itiro would also be the president of the Tokyo University of Education from 1956 to 1962, helping the university progress. With his many leadership

roles, Sin-Itiro was able to influence and inspire many younger scientists in quantum physics, leaving behind a rich legacy.

In the late 1970s, Sin-Itiro received some tragic news. He had esophageal cancer. He did not share his diagnosis with anyone until 1978. According to Matsui's book: "Prof. Tomonaga suddenly told us, 'I am not feeling so well these days. Today I got a little carried away and talked too much. I am tired now.' He then said: 'You should ask things of the old as soon as possible. Otherwise, you may never know the answer, as they will die off one by one.'" On March 14, 1979, Matsui visited Sin-Itiro in the hospital as he was recovering from a tumor removal surgery. The surgery damaged his voice so he could not speak, and communicated with Matsui by writing notes. The two men joked with each other before Ryoko, Sin-Itiro's wife arrived and Matsui left. He was sure he would see Sin-Itiro again. Unfortunately, only a few months later, on July 8, 1979, Sin-Itiro passed away.

After over 40 years of research and work in science policy, he had helped inspire future scientists while also contributing to safer nuclear testing in Japan. As Matsui stated, reflecting on the loss of such an influential genius: "I suppose that no matter how much time elapses, we shall never be able to forget Prof. Tomonaga."

Chapter 6

Claude Elwood Shannon – "The Tinkerer"

The narrow corridors of Bell Telephone Laboratories are relatively silent as most employees have already left for the day. The company's open-door policy seems to extend past office hours, as lights from some of the late shift workers spill into the corridor. However, not all the doors are open. At the end of the hall, one closed door makes the corridor feel cold and unwelcoming. The brass plaque on the door reads: "Claude E. Shannon."

Suddenly the door slowly opens. In a few minutes, many of the workers hear the rattle of a bicycle tire and a 'whoosh' past their door. The newer curious ones poke their heads out while the more seasoned employees already know what the commotion is. As their eyes adjust to the dim light, the newer employees can see a man unicycling and juggling toward them. His concentrated face betrays no feeling of embarrassment or concern, but instead, he is merely focused on the balls in the air, creating perfectly balanced arcs.

***

Claude Shannon's character, as well as his many accomplishments, makes his story worth telling. Not only did he create the field of modern information theory just from his master's thesis, but he also created some of the first devices testing "artificial intelligence." Claude can also easily be categorized as an eccentric genius, constantly making new toys and puzzles out of machine parts. He enjoyed unicycling and juggling and would do both throughout his entire career, not caring if his work colleagues noticed or not. As Claude was quite shy, his reclusive nature only added more mystery to his reputation and legacy, as we'll discuss later. While Claude became a celebrity within the scientific community (giving autographs to fellow researchers in droves, and even doing a photo shoot for *Vogue* magazine), he was relatively unknown in the public sphere and continues to be so. Yet the public owes Claude so much as he made great strides to advance our technology. As writers Jimmy Soni and Rob Goodman explain in *A Mind at Play*: "Before Shannon, information was a telegram, a photograph, a paragraph, a song. After Shannon, information was entirely abstracted into bits." Because of his incredible work and rather unknown legacy, we (like a few other writers before us) feel compelled to tell Claude's story, hoping others will share it in an effort to make him more well-known.

Born in the then-small town of Petoskey, Michigan, on April 30, 1916, Claude found from a young age that he loved tinkering with machines. His parents were Claude Elwood Shannon Sr. and Mabel Catherine Wolf. His father was an entrepreneur while his mother was a teacher. Little is known about Claude's early life, except for a story where he and a friend created a homemade telegraph machine. Using barbed wire strung between houses, Claude and his friend could supposedly send messages to each other in Morse code. This story highlights Claude's unique thought process that he would continue to use throughout his life. He was taught at primary and secondary schools in nearby Gaylord Michigan, where Claude's mother was the principal at his high school. His father served as a local judge for a while. Claude Sr. was much older than an average father and was already sixty-nine when Claude Jr. graduated high school, making their relationship rather distant. Claude was especially interested in science and mathematics and graduated high school in 1932. All through childhood, Claude's hero and role model was inventor Thomas Edison. Claude would later find out that Edison was a distantly related cousin, which any expert might find a bit ironic.

After graduating high school early at age sixteen, Claude began looking at universities. Claude chose the University of Michigan, double majoring in electrical engineering and mathematics. "I wasn't really sure which one I liked best," he would later say. According to a recent study by *The College*

Board, around "25% of college students choose to double major while in college," explained a 2021 article from Bay Atlantic University. This number decreases as students advance through the university system, due to more technical workloads and class requirements. The same study cited that a third of students worried about getting the requirements needed to graduate with both degrees. This requirement may be easier for students who have degrees that complement each other (like Claude's) than two different degrees (like Kenna's undergraduate degrees in English and Biology or Brian's undergraduate in Commerce/ Masters in Finance and AI Strategy program at MIT). Possibly because of the complimentary degrees (and definitely due to his smarts), Claude was able to graduate in 1936, after four years. After graduation, Claude saw a job opening for a master's student and assistant at the Massachusetts Institute of Technology (MIT). Claude later said that "I pushed hard for that job and got it." With an acceptance, Claude moved the farthest away from home he had ever been.

The Early Years at MIT

At MIT, Claude was advised by professor Vannevar Bush. Bush had already gained a reputation for being a skilled engineer and genius. He had created one of the first analog machines called a "differential analyzer" (see chapter 7). Bush could be considered the 20th-century version of *Iron Man*, because he was always creating new machines when he wasn't teaching. According to *A Mind at Play*: "It was Vannevar Bush who brought analog computing to its highest level, a machine for all purposes, a landmark on the way from tool to brain. And it was Claude Shannon who, in a genius accident, helped obsolete it." With Bush, Claude helped improve the differential analyzer, "which used a precisely honed system of shafts, gears, wheels, and disks to solve equations in calculus," stated a 2021 IEEE (Institute of Electrical and Electronics Engineers) article. Claude would continue to cite Bush as an important figure in his life. According to Soni and Goodman, two of the leading experts on Claude Shannon, Bush encourage Claude to generalize his focus and not specialize in one particular field. "As Shannon once explained to Bush: 'I've been working on three different ideas simultaneously and strangely enough, it seems a more productive method than sticking to one problem,'" Soni and Goodman wrote in a 2020 article. This generalization and mentorship from Bush helped Claude to achieve his master's degree and Ph.D. both in 1940. Once Claude was close to finishing his master's degree, Bush advised him and helped him pursue a Ph.D. in mathematics, seeing Claude's clear brilliance in this field. This process was far from easy, as Claude had many hoops to jump through in order to graduate. "To satisfy the language requirement, he had to hire tutors in French and Spanish and worked repeatedly with stacks of flashcards," stated a 2001 *MIT News* article. "He had to take the German exam twice."

To achieve his master's degree, Claude wrote a prize-winning master's thesis (the thesis won the Alfred Nobel Prize, not to be confused with the other, more famous, Nobel Prize, also inspired by Alfred Nobel). According to Soni and Goodman: "At 21, he published what's been called the most important master's thesis of all time, explaining how binary switches could do logic, and laying the foundation for all future digital computers." For his thesis, Claude applied a type of mathematical logic called Boolean logic (named after mathematician George Boole) to binary switches (on or off) within electric circuits. Boolean logic uses true or false (1 or 0) values for digital circuits and gates. Using this logic, Claude could simplify the design of the relays (switches) within the system and even use these setups to solve Boolean algebra problems. With the binary system, Claude could design his switches to work in either a parallel circuit or a series circuit. According to science cartoonist and writer Randall Munroe (also known as XKCD) in his book *What If 2:* "In his 1937 master's thesis, mathematician Claude Shannon showed how vacuum tubes could be arranged to implement any set of logical steps, providing a blueprint for how to build Alan Turing's universal computer by using practical electrical components."

While his master's thesis was 72 pages long, Claude wrote it in a rather dry manner, which, surprisingly, some companies thought would be more accessible to their customers. On a 2002 virtual memorial wall for Claude, a Dr. Penelope Smith wrote that when she was nine: "I bought a primitive 'computer kit' called 'Geniac' by mail. The kit was basically a bunch of rotary switches and simple experiments. However, they bundled Shannon's master's thesis: *On the Symbolic Analysis of Switching Circuits* with it. I read it and decided to become a mathematician. It was such a clear introduction to Boolean methods that a nine-year-old girl could understand it and it laid the foundations for designing computer circuits."

After Claude completed his master's degree, Bush encouraged him to pursue a new field of research outside of his comfort zone, in genetics. Bush recommended genetics as it reflected much of the mathematical theory that Claude had already studied. We don't know how much Claude enjoyed this idea, but we do know he traveled to Cold Spring Harbor to study genetics. His Ph.D. thesis "An Algebra for Theoretical Genetics," was difficult to complete and eventually Claude gave up on following through with it, though he still received his Ph.D.

While studying at MIT, Claude worked summers as an intern at Bell Telephone Laboratories (now Nokia Bell Labs), a large technology company. Bell Labs was created in 1925 when various engineering departments within American Telephone and Telegraph (AT&T) company and the Western Electric Company merged into a group called the Bell Telephone Laboratories. This group was solely dedicated to research, especially in telecommunications and sound systems. According to *A Mind At Play*: "By the time Shannon joined Bell Labs, the curious mix of techniques, talent, culture, and scale had turned the modest R&D wing of the phone company into a powerhouse of discovery."

Bell Labs was home to many Nobel Prize winners, including the winners of the 1947 Nobel Prize for creating the transistor (see chapter 7). As a summer intern, Claude would have no doubt been excited by the new inventions being developed at Bell Labs. His close relationships with scientists in the company would eventually prove beneficial for his career.

After graduating, Claude received a National Research Fellowship and decided to work at the Institute for Advanced Study (see chapter 6) at Princeton. This institute was a hub for bright minds like Claude's, and he had occasional discussions with scientists like Albert Einstein or mathematicians like John von Neumann. Claude even told Arthur Lewbel (now a professor of economics at Boston College), a friend from MIT's juggling club, about one particular interaction with Einstein. According to Lewbel's retelling (via the virtual memorial wall):

> "Claude told me this story. He may have been kidding, but it illustrates both his sense of humor and his delightfully self-deprecating nature, and it certainly could be true. The story is that Claude was in the middle of giving a lecture to mathematicians in Princeton when the down at the back of the room opens and in walks Albert Einstein. Einstein stands listening for a few minutes, whispers something in the ear of someone in the back of the room, and leaves. At the end of the lecture, Claude hurries to the back of the room to find the person that Einstein had whispered to, to find out what the great man had to say about his work. The answer: Einstein had asked directions to the men's room."

He had other interactions with Einstein as well. Claude later said that he passed Einstein early in the morning on the way to the Institute and that Einstein "usually would walk along in sort of bedroom slippers and had old clothes hanging on and he looked like a transient almost…He didn't know really who I was but he'd wave back. Probably he thought I was some kind of weirdo." This story was just one example of the many influential scientists working at the Institute of Advanced Study that Claude would interact with. Unfortunately, his time there would not last, as the US headed off to war.

A War of Codes

In 1941, Claude was approached by Thornton Fry, a mathematician from Bell Labs. Fry was already familiar with Claude's works, seeing him during the previous summers. As Claude later stated: "Fry said: 'Could you join my committee and work on this important subject for the national defense effort?'" The important subject was...anti-aircraft systems. Claude was interested in Fry's offer. In a 1982 oral history interview with Robert Price, Claude said: "It seemed to me that I would be safer working full-time for the war effort, safer against the draft, which I didn't fancy. I was a frail man, as I am now." With his mind decided, Claude resigned from his fellowship and accepted Fry's offer, avoiding the front lines of battle. Fry got him a contract with the National Defense Research Committee (which Vannevar Bush helped lead) to do mathematical analysis. Claude worked with a team of mathematicians on anti-aircraft missiles and their control systems. Anti-aircraft weapons (also called "fire control") worked by targeting incoming planes with missiles and shooting them down before they could fire their ammunition. As Claude told Price, the anti-aircraft weapons "were just sitting there on the edge of the [English] Channel, and the bombs would come over and only a fraction of them ever got to London or places like that." Thanks to Claude's and his team's work, the anti-aircraft systems were highly effective during World War II and saved thousands of lives. These systems were also the important foundation for the anti-missile systems that the US military uses today.

After his work on the anti-aircraft systems, Fry asked Claude to begin working in a different field entirely…cryptography. Throughout his life, Claude loved puzzles and he had been solving cryptograms since he was a young boy. Claude enjoyed the freedom at Bell Labs to explore new methods of cryptography. His work was highly classified at the time, and only later would it be known that he helped in developing the secret communications system that Winston Churchill and Franklin Roosevelt used during the war, which included voice scrambling devices for telephones. Even though Bell Labs had an open-door policy, Claude exempted himself from this culture, as his door was often closed. Though he was shy and reclusive, he also was happy to discuss any sort of project with his fellow colleagues.

In early 1943, Claude was asked to meet a new visitor at Bell Labs, scientist Alan Turing. Turing is most remembered for solving the "enigma machine", the device that encoded messages for the Nazis. In 1943, Turing arrived in the U.S. to gather information on current cryptography methods and to ensure that the Bell Labs device for encrypted communication would be secure enough for the U.K. to use. As *A Mind at Play* stated: "Dr. Turing would have to give his stamp of approval that the system was unbreakable." He and Claude became friends rather quickly. "Turing and I had an awful lot in common..." Claude later told Price. "We spent much time discussing what's in the human brain. How is the human brain built, how it works, what can be done with the machine, and whether you can do anything with the machine that you can do with the human brain." While Turing could not tell Claude what he was working on (and the same went for Claude), Claude had some ideas. They did swap many theories, however, especially on computing. "I think Turing had a great mind, a very great mind," Claude said. "And some of his concepts were terrific [like] the universal machine." Later, in the 1950s, Claude would reunite with Turing in the U.K., where they both discussed electronic chess-playing machines. That would be the last time Claude saw Turing before Turing's tragic suicide.

In 1945, Claude compiled his many ideas and advancements in cryptography into a highly classified paper: *The Mathematical Theory of Cryptography.* The paper analyzed the current "unbreakable" encryption methods. Claude's paper also "made the astute observation that language, particularly English, was redundant and predictable," explained the Nokia Bell Labs archive. Additionally, the paper "proved that in a point-to-point error-free link where an eavesdropper observes the transmitted message with no error, with enough common randomness between the transmitter and the receiver, a perfectly sure communication link is achievable."

Claude's theories on cryptography are still utilized by those working on quantum encryption. Because quantum computers promise to factor large numbers at a faster rate, many experts worry that this technology will be devastatingly effective in decrypting our current digital encryption systems. For this reason, many quantum companies, as well as researchers, are developing ways to "quantum-proof" encryption, often by using quantum computing itself. As discussed in chapter 1, algorithms like Grover's and Shor's could be used to decrypt current classical encryption methods. Using these algorithms as templates, others are working on ways to overcome these challenges with more secure methods of encryption. Though Claude would have no idea that his work in cryptography would be so influential, it would be his next big paper that would revolutionize the digital era.

A Dive into Information Theory

Both Claude's master's thesis and his work on cryptography helped influence his 1948 paper *A Mathematical Theory of Communication* in the *Bell System Technical Journal* which some experts believe took a decade to produce. Many of the paper's early work went onto napkins that Claude had scattered about his apartment. The paper helped create the field of information theory and sparked a firestorm of interest in digital communications. According to Ioan James, a biographer for the Royal Society: "So wide were its repercussions that the theory was described as one of humanity's proudest and rarest creations, a general scientific theory that could profoundly and rapidly alter humanity's view of the world." The paper launched Claude's celebrity status almost overnight, leading to his later title of "Father of the Information Age." In its first year of publication, the paper (published as a book) sold around 6,000 copies, and by 1990, it sold more than 51,000 copies. According to *A Mind at Play*: "In time, the thoughts developed in these seventy-seven pages in the *Bell System Technical Journal* would give rise to a digital world: satellites speaking to earth in binary code, discs that would play music through smudges and scratches…the world's information distilled into a black rectangle two inches across." All this came from Claude's theory. He was thirty-two when he wrote the paper.

Claude's theory set out to answer two big questions: how can messages be encoded in a noise-free environment and how can messages be reliably transmitted in the presence of noise? To answer these questions, Claude thought about a communications channel like a flowing river. The river could only contain so much water (the information) before it overflowed. This led to Claude realizing that communication channels had a limit to how much information they could send (what we now call the bandwidth or also called the "Shannon Limit"). Claude argued that "any information imaginable can be quantified in the same way, regardless of the medium of communication," explained a

2022 *Computer-History* article. This meant that all information could be reduced to digital values, known as binary digits or "bits" (which Claude credited fellow Bell Labs mathematician John Turkey for coining, after proposing "binit" and "bigit"). A channel's bandwidth or capacity was measured in "bits per second." As Bell Labs scientists John Pierce and Michael A. Noll later explained in their book *Signals*: "He proved that if the information rate of a message source is less than the capacity of a noisy chancel, messages from the source can be transmitted without error (strictly, with fewer errors than any assigned number, however small)." This helped to answer both original questions.

To further answer these questions, Claude focused on the noise within the channel. He found that the noise in the channel was a challenge to the "entropy" of the channel, a word he borrowed from physics (possibly inspired by physicist John von Neumann at the Institute of Advanced Study). "In the most basic terms, Shannon's information entropy is the number of binary digits required to encode a message," stated a 2002 *Scientific American* article. Depending on the noise of the channel, the entropy from the information's source could be affected. To overcome the noise "Shannon demonstrated that you could deliver this digitized information over communication channels containing more noise by using probability theory to quantify the amount of noise and then using redundancy to eliminate it" added the *Computer-History* article. As a mathematician, Claude realized that communication included many different probabilities. To better the chances of a message being sent successfully through a noisy channel, he proposed redundancy for the message. Redundancy happens in multiple ways, such as repeating a message multiple times (Hello? Hello?) or having multiple letters to correct any message errors (heplo is clearly a typo for hello and we see this due to the redundancy of the letter l). Referring to the image of Claude's theory in the photo insert, you'll see that Claude has minimized the idea of communication to its basic variables. "You speak into a phone (source); the phone encodes the sound pressure of your voice into an electric signal (transmitter); the signal passes into a wire (channel); a signal in a nearby wire interferes with it (noise); the signal is decoded back into sound (receiver); the sound reaches the ear at the other end (destination)," explained *A Mind at Play*.

Combining the ideas of channel capacity and redundancy, Claude changed how people thought of communication. As bestselling author, quantum physicist, and mathematician Chris Ferrie told Kenna in an interview: "It is a theoretical framework for understanding the process of transmitting information from one point to another. The key ideas are that information can be encoded in various ways and that it can be transmitted over noisy channels. The theory has been hugely influential in the development of digital communication systems, and it continues to be relevant to researchers today as they try to design ever more efficient communications systems."

After publishing his paper, Claude became a well-known figure in the scientific community. "I didn't think in the first stages that it was going to have a great deal of impact," Claude later said. Throughout his life, Claude cared little about publicity or awards. In fact, he tried to avoid the spotlight as much as possible and cringed when people compared him to Einstein (which would happen for the rest of his life). However, it was difficult to do so with his new theory becoming so widely adopted. According to a 2016 *New Yorker* article: "Almost immediately, the bit became a sensation: scientists tried to measure birdsong with bits, and human speech, and nerve impulses." The wide use of the "bit" bothered Claude to the point where he wrote a paper called "The Bandwagon" in 1956, discouraging this type of use. However, historians found that these widespread uses for the "bit" were actually appropriate, as the theory could accurately be applied to fields like biology or social networking to give deeper insights. "It made quite a splash," explained Dr. David Thomson of Queens University. "It was not really believed by many when it first came out, it seemed so unique. Even some of the higher-ups at Bell Labs didn't know what to make of it." Claude became busier than ever after publishing his ground-breaking paper, not because of his work, but because he had fallen in love.

An Intellectual Partner

In late 1948, Claude met Mary Elizabeth "Betty" Moore at Bell Labs. Claude had already been married and divorced by this point. In 1940 he married Norma Levor, an intellectual in many elite social circles, after only a few months of dating. However, the move to Princeton's Institute of Advanced Study drew them apart and they divorced only a year later, with Levor marrying a screenwriter in Hollywood sometime after. As a new bachelor, Claude dated a few women while at both the Institute of Advanced Study and Bell Labs, but ultimately found himself to be rather lonely. When he met Betty in 1948, he summoned the courage to ask her out to dinner. Soon they were dining together every night. They spent time at each other's apartments, discussing mathematics and playing music together, as he played clarinet and she, piano. Claude's rising celebrity did little to hinder their relationship as Claude was completely infatuated with Betty.

In Betty, Claude found his intellectual equal. Betty Moore Shannon is another unsung hero of computer science, as she was a genius in her own right (hopefully some future book focuses on her story). She excelled in high school and then attended the New Jersey College for Women (now Rutgers University), achieving a degree in mathematics. The very next day after her graduation, Betty began her job at Bell Labs as a "computer" (see chapter 7). According to a 2017 IEEE obituary for Betty: "She was promoted to Technical Assistant at Bell, and worked with John Pierce, the inventor of the communications satellite..." She also published a Bell Labs Technical Memorandum called "Composing music by a Stochastic Process." After meeting Claude, Betty quickly became smitten with him, as he did with her. They married rather quickly, on March 22, 1949, and would remain devoted to each other for the rest of their lives. They would go on to have three children together, starting in 1951 when Betty quit her job at Bell Labs to raise the family.

Betty was exactly the partner Claude needed. She had his same wry sense of humor, and would often be the first person to hear about his new ideas. She would also edit his many papers. Betty would later say that "some of his early papers and later papers are in my handwriting...and not in his, which confused people at first." Historians credit Betty with helping to "fill in" some of Claude's later work, as he would often jump ahead. Claude would later state: "I think I'm more visual than symbolic. I try to get the feeling of what's going on. Equations come later." Betty also helped push Claude out of his comfort zone by accepting speaking invitations to events he would have rather avoided. Betty was also the perfect intellectual partner for Claude because she supported his hobbies of unicycling and tinkering.

Tinkering and Other Hobbies

Thanks to his lifelong fascination with games and puzzles, Claude had become quite a mechanical tinkerer. Much of his tinkering went into creating mechanical toys, such as a juggling machine, an automatic Rubik's cube, a rocket-powered Frisbee, and a trumpet that shot out flames when someone blew it (the last two sound especially worrisome). His love of tinkering even inspired the "Toymaker" character in the original 1982 movie *Bladerunner*.

While Claude created many of his mechanical devices at Bell Labs, his home (which he called "Entropy House") was ground zero for his tinkering, to the point where it was easier to find things he hadn't tinkered with than things he had. According to mathematician Edward Thorp, remarking on a 1961 visit to Claude's house (in which he was helping Claude to make a wearable computer), the house was "a gadgeteer's paradise. There were hundreds of mechanical and electrical categories, such as motors, transistors, switches, pulleys, gears, condensers, transformers, and so on."

As the 2002 memorial wall explained, Claude was also working on inventing the first motorhome by installing a shower in his Volkswagen microbus. Claude had even created a rotating false wall in his home, as well as "a gadget whose only purpose when it was turned on was to open it up, release a mechanical hand, and turn itself off," explained Soni and Goodman. While it might seem like an embarrassment for a successful scientist like Claude to focus on toys, he didn't care enough and was too focused to change his hobby. His MIT colleague Robert Gallager later stated: “Claude Shannon was a very gentle person who believed in each person’s right to follow his or her own path.”

Besides tinkering, Claude was an avid unicyclist. Betty bought Claude his first unicycle while they were dating, and by the end of his life, his house in Massachusetts had over 30 unicycles. In the evenings, Claude would ride his unicycles up and down the corridors of Bell Labs, as well as MIT later on. His habit both confused and worried his colleagues. “When he returned to MIT in 1958, he continued to threaten corridor walkers on his unicycle, sometimes augmenting the hazard by juggling,” stated a 2003 article from the University of St. Andrews. “No one was ever sure whether these activities were part of some new breakthrough or whether he had just found them amusing. He worked, for example, on a motorized pogo stick, which he claimed would mean he could abandon the unicycles so feared by his colleagues.” Claude also apparently created a two-seater unicycle, but few tested it with him as most were too frightened to give it a try.

As if tinkering and unicycling were not enough, Claude also enjoyed juggling. He took a scientific approach to juggle, as he even tried to write a mathematical paper on juggling for *Scientific American*. Lewbel saw Claude’s scientific approach firsthand when he arrived at Claude’s house one day. “Claude said to me: ‘Do you mind if you hang upside down by your legs?’” Lewbel recalled on the virtual memorial wall. “He [Claude] had realized that while bounce juggling is much easier than toss juggling in terms of energy requirements, throwing upwards as in toss juggling is physiologically easier, and so he wanted to try combining the two, which meant bounce juggling while hanging upside down.” Claude often used juggling as a way to shorten his speeches at events, making himself feel more comfortable with being in the spotlight.

Theseus the Mouse and Testing Machine Intelligence

In 1950, one of Claude’s tinkering projects became quite a sensation. Betty had bought Claude an Erector set (a metal construction set) which Claude used to construct an electronic mouse. As Betty told an interviewer later: "I went out and got him the biggest Erector set you could buy in America. It was fifty bucks and everyone thought I was insane! Giving it to a grown man! But the fact of the matter is that it was extremely useful and I used it to try out different things." Betty did most of the wiring for the electric mouse while Claude designed a special maze for his new device. In calling the mouse Theseus, Claude designed a type of "intelligent" system for his electronic mouse to use to escape a maze, similar to the Greek myth of Theseus. "Shannon created a system of mazes for his mouse that he could modify at will," the *Computer-History* article explained. "He designed the mouse's relay circuit brain to travel through the corridors of his mazes until it discovered the exit. Once a given maze had been plotted, he could put the mouse in any location it had been in before, and it would use its prior experience to immediately plot a path to the exit." This test revealed Theseus' machine intelligence (and you might notice a similar process with the "Roomba" or other floor vacuums that "learn" in a likewise manner, as both the "Roomba" and Theseus had sensors to monitor where walls are). Claude kept testing Theseus by changing his mazes and Theseus would continue to add new pathways to its memory storage. In this way, Theseus became one of the first devices to demonstrate "artificial intelligence." Bell Labs filmed some of these tests and aired them publicly, adding more to Claude's celebrity status.

Claude also tested machine intelligence using chess. "An avid chess player, Professor Shannon built a chess-playing computer years before IBM'S Deep Blue came along," the 2001 *MIT News* article said. This machine, according to Soni and Goodman, had over 150 relay switches and took around 10 to 15 seconds to decide a move. "While on a trip to Russia in 1965, he challenged world champion Mikhail Botvinnik to a match. He lost in 42 moves, considered an excellent showing." Claude's fascination with chess resulted in a 1950 paper titled *Programming a Computer for Playing Chess* in the *Bell System Technical Journal*. In the paper, Claude discussed how early modern computers could play against human opponents. "The paper estimated that the number of possible chess games was somewhere around 10 to the 120th power, a number larger than the number of atoms in our universe," the *Computer-History* article added. "The idea, now known as 'Shannon's Number' implies that a brute-force algorithm wouldn't be a plausible way of teaching a computer to play chess."

The Later Years at MIT

In 1956, Claude transitioned from industry into academia and became a professor at MIT, in 1957. As one of MIT's most famous alumni, Claude's lectures drew large crowds. He continued to consult for Bell Labs for many years, but moved into a professorship to, as many experts posit, avoid the spotlight. At MIT, Claude continued to work on machine intelligence. He once stated: "my mind wanders around and I conceive of different things day and night. Like a science fiction writer, I'm thinking, *what if it were like this?*" He advised some students, but not very many, as many were afraid to approach him with ideas for future research. He also admitted to others that he did not enjoy advising students' theses. According to Thomson, who worked at Bell Labs during Claude's consultancy phase, Claude gave his approval to a mathematical solution that Thomson's friend Michael Noll had found for an information theory class taught by a Bell Labs scientist. As Thomson explained, "[Noll] approached this scientist with his solution and said, 'It's approved, Shannon says so.'" This was just one example of Claude's willingness to discuss different scientific and mathematical problems with interested students, even at different facilities, though his usual way was asking questions instead of answering them.

During the later years of his life, Claude and Betty became interested in the stock market. It started with Betty becoming intrigued, and quickly Claude was drawn in. While they were not motivated to make gains from analyzing the market, they did enjoy noticing patterns in the stock fluctuations. Their interest in the stock market led to Claude being able to enter the ground floor of some emerging Silicon Valley companies that would become huge later on, including Hewlett-Packard. Claude was also able to help advise these companies, giving them valuable wisdom found nowhere else. These early investments, mixed with their analysis, made Claude and Betty a decent amount of money. Claude even gave a lecture on the stock market at MIT in the University's biggest lecture hall. Even with the bigger space, late attendees found it to be "standing room only."

Throughout his character, Claude participated in many organizations. He was a member of the National Academy of Sciences, the American Philosophical Society, the Royal Society of London, and the Leopoldina Academy. He also won many awards, including (but not limited to): the National Medal of Science, the Stuart Ballantine Medal, the Harold Pendant Award, the John Fritz Medal, and the Harvey Prize (of which he was the first recipient). He was also the first recipient of the Shannon award in 1975 by the IEEE, for work in information theory and electrical engineering. In 1985, Claude was also the first recipient of the Kyoto Prize for his work on information theory. "Known as the Japanese Nobel Prize, the Kyoto Prize awards people who have contributed significantly towards progress in science, technology, art, and philosophy," explained the *Computer-History* article. Claude may also be the winner of having the most honorary degrees from other universities, including the University of Oxford, the University of Edinburgh, Princeton University, Yale University, Carnegie-Mellon University, the University of Michigan, the University of East Anglia, the University of Pennsylvania, Tufts University, and Northwestern University. According to Soni and Goodman: "He accumulated so many honorary degrees that he hung the doctoral hoods from a device that resembled a rotating tie rack (which he built with his own hands)." Yet Claude thought very little of these awards and degrees "I don't think I was ever motivated by the notion of winning prizes, though I have a couple of dozen of them in the other room," Claude later said. "I was more motivated by curiosity. Never by the desire for financial gain. I just wondered how things were put together or what laws or rules govern a situation, or if there are theorems about what one can't or can do. Mainly because I wanted to know myself."

A Tinkerer's Legacy

Claude retired from MIT in 1978, after having a full and successful career. In his later years, he struggled with

Alzheimer's disease. The diagnosis first came in 1983, as Claude began forgetting more and more important faces and dates. "The communications channel to one's memories—one's past and one's very personality—is progressively degraded until every effort at error correction is overwhelmed and no meaningful signal can pass through," explained the 2002 *Scientific American* article. "The bandwidth falls to zero." Betty made sure to keep Claude's condition private and took care of Claude until 1993 when Betty entered an assisted living center. Claude continued tinkering till the end of his life, even working on his own walker. He passed away on February 24, 2001. Betty would pass away some years later in 2017. Lewbel remembered his own final moments with Claude, writing on the 2002 memorial wall: "As sad as it is to see anyone's light slowly fade, it is an especially cruel fate to be suffered by a genius... Despite the loss of memory and reason, he was every bit as warm, friendly, and cheerful as the first time I met him." Others echoed similar sentiments, mourning the passing of whom Nokia Bell Labs called "a Goliath among giants."

Claude's work impacted many other scientists, including Artur Ekert, a Professor of Quantum Physics at Oxford University and a pioneer in the field of quantum mechanics (and with whom Brian shared a stage at a 2022 Madrid conference). Best known for his 1991 paper "Quantum Cryptography Based on Bell's Theorem," in which he proposed a method for secure communication using the principles of quantum mechanics, Ekert's work led to the development of quantum key distribution; a method for generating and sharing secret keys between two parties. Claude's and Ekert's work have many similarities in that both of them revolutionized the way we think about communication and information.

While Claude Shannon's story is full of brilliance and fame (as he made the covers of *Life* and *Time* magazine, as well as had a photoshoot done by Henri Cartier-Bresson for *Vogue* magazine), his story is still fairly unfamiliar in the public sphere. Yet without his work, the digital devices we take for granted (or curse at sometimes when they lag) wouldn't exist. As writer James Gleick explained: "It's Shannon whose fingerprints are on every electronic device we own, every computer screen we gaze into, every means of digital communication." Not only did Claude make enormous strides in technology and engineering, but he also was not afraid to be himself. According to *A Mind at Play*: "Shannon's admirers are just as quick to compare him to M.C. Escher or Lewis Carroll as they are to put him in the company of Albert Einstein or Isaac Newton." His eccentricities made him more memorable and loved by those around him, and for the many of us who didn't know him personally, they make him more human.

Chapter 7

Frances "Betty" Holberton – "The Lady of Logic"

The room is as hot as it is silent as everyone holds their breath. At the back of the room, behind the rows of journalists, government officials, and scientists, Betty Holberton excitedly grasps her friend Jean Bartik's hand. On the other side of her, fellow female programmer Kay McNulty taps her foot impatiently, waiting for the demonstration to begin.

"You ran through everything this morning," Jean whispers to Betty. "I promise it'll be fine, you do more programming than anyone here."

"Oh, I know that," Betty responds. "It will definitely work."

One of the journalists shushes the women as the giant machine at the front of the room begins to whirr. The following seconds are filled with loud clanking noises and flashing lights until finally, the symphony of technology ends as an IBM punch card slides into its tray. The demonstrator gingerly picks up the card and holds it high in the air.

"Gentleman and ladies," he says triumphantly. "It works!"

***

You may not realize it, given the current low percentage of women in the technology sectors, but computer science has a history filled with female pioneers. Only recently have many of these stories come to light, including Betty Holberton's. For nearly fifty years, her story and those of the rest of the "ENIAC 6," the first modern computer programmers (who happened to be women) were untold. It wasn't until the mid-1990s that a female lawyer and computer scientist named Kathy Kleiman found the "ENIAC 6" and compiled their stories into a recently published book called *Proving Ground*. Before Kleiman's work, these women were simply referred to as "refrigerator ladies," a term for female models who helped sell electronic devices like desktop calculators or computers like the ENIAC (which stood for Electronic Numerical Integrator and Computer). Kleiman saw a photo of the ENIAC 6 and realized that instead of modeling, these women were actually programming the machines. Thanks to Kleiman's thorough research and dedicated storytelling, these six women will hopefully now be remembered for their programming expertise, and their efforts to pave the way for the numerous women to follow within computer science.

There are many reasons we picked Betty specifically as one of our "giants." From being a brilliant programmer to focusing on women's rights in this field, Betty's monumental career in computer programming lasted for over 40 years, as she retired from the National Bureau of Standards, NBS (now known as the National Institute of Standards and Technology or NIST) in 1983. During that time, she helped standardize some of the original codes needed for using early commercial computers, such as C-10, COBOL (Common Business Oriented Language), and Fortran (Formula translation). She worked very closely with Grace Hopper, arguably one of history's most well-known women in computer science.

As you'll see from Betty's story, she focused largely on the user experience of computing, making sure the technology was accessible to anyone. She also focused on making sure she and her female colleagues were treated fairly, especially during the time of rampant sexism in the 1950s and 1960s. Betty's generation of the ENIAC 6 came a couple of decades before the brave women at NASA (showcased in the movie *Hidden Figures*) and used their intelligence and talents to fight against a male-favored system. Because of their bravery and advancements, the stories of Betty and her fellow programmers (Jean Bartik, Kay McNulty, Fran Bilas, Marylin Meltzer. and Ruth Teitelbaum) deserve to be told over and over again.

While Betty and her team did not work directly on quantum technology, their coding and programming expertise laid the technical foundation needed for today's scientists and engineers to build a quantum computer. Indeed, many quantum companies are currently developing "hybrid" computing systems that combine classical and quantum computing to solve complex and difficult problems. Commercial quantum computers are still being constructed, similar to the early days of the ENIAC computer, where few know what these machines may be able to do. Perhaps we'll see a similar story to Betty's, where fearless and brilliant quantum programmers help to create the next generation of computers. Even while this is in progress, Betty still plays an extremely important part in our society, as a role model and inspiration for all of us, not merely female programmers, to use our creativity and aspirations to build something that may help others around us and better our future.

Born on March 7, 1917, as Frances "Betty" Snyder, Betty grew up in Narberth, Pennsylvania, the third of eight children. Though the town of Narberth was more rural and remote, this didn't stop Betty or her family from learning more about the world around them. Thanks to inheriting books from an English-teaching uncle, Betty had no shortage of books to read, especially literary classics.

Betty actually was born into a rich scientific lineage. Her grandfather, Monroe Snyder, graduated from the University of Michigan in 1872 (about 45 years or so before Elmer Imes in chapter 3) and taught mathematics and astronomy at Narberth's Central High School for boys (the School would become co-ed in 1983, at least 150 years after it was founded). Monroe also was the Director of the local observatory and a member of the Royal Astronomical Society. He also made his own significant contributions to science, as he suggested founding the NBS originally. He did this to ensure "the products of electrical instrument makers and manufacturers were precise," stated *Proving Ground.*

Thanks to Monroe Snyders' dedication and active circle of intelligent friends (including Alexander Graham Bell and Thomas Edison) the NBS was established in 1901 and became the US government's first research laboratory dedicated to physical sciences. As Betty elaborated in a 1983 interview with James Ross of the Charles Babbage Institute: "In 1884, he [Monroe] had a conference set up of international, what they called electricians, they were physicists, so he could deliver this paper on establishing the National Bureau of Standards, and it became part of the document that went in with the declaration of it." Being the secretary of the United States Electrical Commission, Monroe was able to help influence future steps toward creating the NBS.

In Monroe's later years, he lived with Betty's family from 1926 to 1932. During that time, he tried his best to impart his scientific wisdom to Betty, taking her for walks around Narberth's main square and discussing all different scientific fields from chemistry to astronomy. During these walks, he would also give Betty money to buy herself a treat, which she used to get an ice cream cone and a pickle, mixing the sweet and sour, as Kleiman wrote in her book.

Because Betty was born cross-eyed and flat-footed, she needed expensive equipment to correct both issues. She reasoned that her parents spent more time with her than her other siblings because of her problems (she was also left-handed, which in the early 1900s, would have been "corrected" to right-hand dominance). Because of this extra attention, Betty did her best to make it up to her parents by "working harder," according to *Proving Ground.* Betty had a different way of thinking than most children her age. At age 15, Betty and her younger brother took apart Monroe's Marchant desk calculator. This calculator was a big clunky machine. As Betty explained in the Ross interview:

> "[We] decided that the divide circuit wasn't working one day, and we took the whole thing apart to see if we could fix it... I labeled every part and diagrammed where it was going to be, we laid it all out on the table, and we got it back together again, but it still didn't work. There was something that wasn't there that we didn't know about, but I had a feel for that kind of thing. I think that is the way that I think."

Doing this exercise would have been rare for a teenage girl in the 1930s during the Great Depression, but Betty's drive to learn was insatiable. "I think kind of like a radar screen," she elaborated in the Ross interview, "On any problem I have, I go around the problem until I have all the facets of it well in my, mind before I...sit down to write anything out on how it's going to be solved." This type of logical reasoning made Betty perfect for her later career in computer programming.

Betty and her family were devout Quakers. As Priscilla Holberton, Betty's daughter, told Kenna in an interview: "Her whole family was Quaker. They were convinced Quaker, not birthright Quaker." This meant that Betty's family was not born into the religion by having Quaker ancestors, but instead, her family joined it willingly, or was "convinced." Quakerism is a sect of Christianity that believes that individuals can experience the divine nature of God for themselves directly. This in turn drives Quakers to try and better themselves and others through meditation, friendship, and giving. Quakers are known for their kind spirits and acceptance of others from any background. As a Quaker family, the Snyders would have worked to uphold these ideals. Betty's father, John Snyder, followed in Monroe's footsteps and taught at Central High School. John was a dedicated speaker. He often recited poems for his family and spoke at different Quaker establishments. His speeches were so well received that he earned a scholarship for one of his children to attend the private Quaker boarding and high school, George School. Neither of Betty's older siblings wanted the scholarship so she took it.

At George School, Betty took co-ed classes and learned a variety of languages and different fields of mathematics. Math was a particular favorite of Betty's and according to Kleiman, "in her senior year [she] took a side job correcting geometry papers for a teacher for a grand sum of $100, which she considered a fortune." With a prestigious high school education, Betty received two college offers, one from the University of Pennsylvania and the other from Temple University. This was almost unheard of for women in the mid-1930s. According to a 2022 report by Statista, only 3.8% of US women completed a four-year university in 1940 (for reference, today it's around 34.1%). This low percentage was due, in part, to general society believing women didn't need a college education and were instead destined for marriage and motherhood. It was also due to lower college participation in general, as few people saw a reason to attend a four-year university (the Statistica report showed around 5.5% of men graduating from a university in

1940). To Betty, these offers must have seemed irresistible. She ended up choosing the University of Pennsylvania as her older brother attended there and he could drive her to and from school.

In 1935, Betty began her courses. Thankfully for her, University of Pennsylvania's College of Liberal Arts for Women had been founded only two years earlier. The College was founded and admitted women students only. According to Penn Libraries Archives "for the first time in Penn's history, women were offered a full-time, four-year, liberal arts, undergraduate degree program. It should be noted, however, that the standing faculty of the College for Women did not include any women."

Even though Penn was co-ed, the classes were still divided by gender, which surprised Betty. The faculty, as mentioned, was still a male majority, something that caused a huge problem for Betty as she majored in mathematics. Some of her math classes were taught by an old Russian professor, who flat-out told the women in his class, "you should be raising children," as Betty said to Ross in the 1983 interview. While Betty suffered through the class, she, unfortunately, found out that he would be her primary instructor for other future math classes. Dissatisfied with this situation, Betty quickly switched her major to journalism. As she told Ross, "this would allow me to take something in every discipline in which they would expect you to know something. I didn't know what I wanted to do. I knew I didn't want to be a teacher." Betty had experienced sexism firsthand, and it, unfortunately, would not be the last time this happened.

With her intense passion for learning, Betty took the maximum number of credits in journalism and graduated in four years taking five years of classes. After her graduation in 1939, Betty got a job at *Farm Journal*, the popular agricultural magazine. "I'll get into the statistical department and do surveys and things like that and get into the writing end," Betty said to Ross about working at *Farm Journal*. Her logical thinking helped save the *Journal* from a costly mistake. As Betty recalled to Ross, one year, the *Journal* ran a survey to see the popularity of selling lipstick to the wives of local farmers. The results showed higher than normal sales. Puzzled, Betty found that the results of the survey were inaccurate due to a mistyped number on one of the desk calculators. She would later use her findings in developing the typing machine for the UNIVAC, the first universal computer, to avoid mistypes by its programmers.

With World War II in full stride, Betty knew that she wanted to do her part. She originally hoped to join the Women Accepted for Volunteer Emergency Service (WAVES) but was denied due to her eye and feet issues. Instead, Betty saw a notice from the Department of Labor Women's Bureau notice that read (as quoted in *Proving Ground*): "[For] war opportunities in science and engineering, you will find that the slogan there as elsewhere is 'women wanted!'" The notice led her to the Moore School at the University of Pennsylvania, where the Army (more specifically the Ballistic Research Laboratory or BRL) was looking for women to help compute trajectories for guns and cannons. Given her background and enthusiasm for math, Betty was quickly hired. At age 25, on August 19, 1942, Betty started her new job at the University of Pennsylvania's Moore School of Electrical Engineering, beginning her journey into computer science.

Time at the Moore School

While World War II was absolutely catastrophic, killing millions, some may argue it also significantly advanced science. With countries collaborating to create weapons using chemistry, physics, or biology, many different scientific discoveries and theories came about during this time. The desperation of these countries including the U.S., also gave researchers almost a *carte blanch* when it came to funding for war research. In the case of the ENIAC, the total was around $500,000 or $7 million today to build the world's first modern computer. Yet the U.S. army paid for this project, and many others, hoping it would give them the advantage they needed to win the war. This wouldn't be the only costly project that the U.S. government would fund, as the Manhattan project cost 4,000 times that of the ENIAC computer.

World War II also helped many women get paying jobs and enter into STEM (Science, Technology, Engineering, and Math) fields more easily. With able-bodied men, including scientists and researchers, being called to the frontlines, the women left behind made sure to fill these vacancies. This included the women at the Moore School. While many women there didn't have college degrees, their interest in science combined with the backdrop of the war allowed them to do a job that they wouldn't have been able to do otherwise. It might even be fair to say that without the desperation caused by World War II, the first modern computer programmers may not have been women, and this story would have been very different than it is today.

At the Moore School, Betty was hired on as an "Assistant Computer," which made a decent wage for a woman at the time. According to an article by NASA on early human computers, the average salary for a junior computer employee was around $1,440 per year or $30,621.70 in today's economy. For many women trying to find work after the Great Depression, this well-paying job seemed too good to pass up. In this role, Betty and the other women would use desk calculators to find specific trajectories for different weapons.

Since the U.S. military lacked the time and expertise to calculate the angle they needed for specific shots, they used "firing tables" to do the work for them. According to *ENIAC in Action:* "the firing tables supplied with each gun specified the elevation required to fire a shell of a certain type at a target a certain distance away and supplementary correction tables allowed a gunman to compensate for the temperature of the gunpowder, the cant of the carriage's axel and other variables." These firing tables were the predecessor for automatic firing systems on missiles and other weapons that are used today. Thanks to the women's work in creating these tables, the U.S. military had effective weapons accuracy.

To calculate these trajectories, the women used desk calculators, which cost thousands of dollars in the 1940s. Unlike the calculating machine of Gottfried Leibniz's time (see chapter 2), these calculators ran on electricity, though hand cranking was still involved. The machines were finicky and could shut down halfway through a calculation. This left lots of room for human error, which had to be quickly corrected. As the U.S. military was spread across the globe during the war, these tables needed to be calculated as fast as possible to match troop movement. This demanded long hours for the computers, as women like Betty would work in two shifts, a day shift from 8 am to 4:30 pm and then a night shift from 4:30 pm to 1 am. The rooms they worked in had no air conditioning, and nearby construction made it noisy on occasion. But still, the women pushed on, making quiet success within the Moore School.

The Moore School was a unique location to host the ENIAC, as it was semi-independent of the University of Pennsylvania. "The school had its own endowment and therefore was actually capable of operating with a great deal of independence from the rest of the University," explained ENIAC inventor J. Presper Eckert in a 1977 interview. With more freedom, the Moore School would become the perfect place to develop the world's first modern computer. The computing program was headed by Princeton mathematician and computer scientist Herman Goldstine. While Goldstine had a difficult time transitioning to the job from academia, his wife Adele, had a different experience. As a mathematician in her own right, Adele taught many of the new recruits and oversaw the program's daily shifts. The computers also had another supervisor who would help: John Vaughn Holberton, who would later marry Betty. Holberton was a Senior Computer in ballistic computations and was a friendly face for the women struggling to learn new mathematical equations.

There are many arguments as to why women specifically were asked to be computers, even after men returned from World War II. Most sources argue that the clerical work of computing made it perfect for women. According to a 1977 interview with Goldstine, "the computer did the dull work. To do a mathematical integration or numerical calculation of any sort is inordinately dull. It's like doing your income tax every day of your life, from morning to night." It was the clerical aspect of early computation that made it seem like it was too lowly for a man to do. As Goldstine added: "that was an era in which there were jobs that were men's and jobs that were women's." While this might fit into the discriminatory landscape of the 1940s and 1950s, the monotony of early computation allowed women to take the stage in a series of monumental "firsts" when it came to computing. Without the drudgery of computing for the Moore School, the amazing women of the ENIAC 6 may never have become what they are today.

A few months after the Moore School program began, ideas for the ENIAC began to emerge. The original idea came from John Mauchly, a physicist. In 1941, Mauchly found himself at the Moore School for a Defense Training Course for Electronics. There he met engineer John Adam Presper Eckert, or J. Presper Eckert, who would help Mauchly design and build the ENIAC.

Eckert was enrolled in the Moore School of Electrical Engineering and had already applied for his first patent at age 21. Together, the physicist and engineer would transform the future of computers. Mauchly's ideas for the ENIAC depicted a machine that could run calculations, in mere seconds, for things like ballistics or even weather forecasting. As Mauchly had worked at the NBS in the summers, examining weather data, he was hopeful that this machine could make the job easier.

The ENIAC's main component would be vacuum tubes. "Vacuum tubes had evolved from incandescent lightbulbs—glass bubbles in which delicate metal filaments shielded from the atmosphere could conduct electricity without burning out," explained *ENIAC in Action.* "But whereas light bulbs were intended simply to turn current into visible light, vacuum tubes manipulated current in a variety of ways. Diodes conducted in one direction only. Triodes have three terminals and were the basic building blocks of amplifiers." The ENIAC would end up using triode vacuum tubes for its basic components, as these components worked like switches or transistors in the machine. Because Philadelphia had a thriving local radio community which created a large demand for these tubes, the triode vacuum tubes could be mass-produced, which would be helpful later on for creating and fixing the ENIAC. According to science cartoonist and writer Randall Munroe (also known as XKCD) in his book *What If 2:* "Transistors and vacuum tubes use different mechanisms to do the same basic task: If they receive an electrical signal, they flip a switch one way, and if they don't, they flip it the other way."

Besides ballistics or weather data, a computer like the ENIAC would make a significant difference to scientists and mathematicians. Before computers like the ENIAC, calculations were done by hand or on calculating machines. The solutions, which may involve tricky math functions, would be included in specific "solution table" sheets, where someone could use the sheet to look up an answer for a particular function without having to do the math themselves. The ENIAC would hopefully remove the need for solution tables as it "was designed with hardware called function tables, capable of storing such tabular data for use in firing table generation," explained a 1996 article for the Institute of Electrical and Electronics Engineers (IEEE).

Since the Moore School Computers were doing tricky calculus equations to solve for ballistic trajectories, the ENIAC's function tables could save significant time and energy in the process. Mauchly had already used vacuum tubes, specifically neon vacuum tubes, to build an electronic counting device. Now he hoped to use the same technology to build the first electronic computer. He pitched the idea to the BRL to help with their ballistics data, and they agreed.

According to the educational film *Computer History: ENIAC Computer History,* the construction contract was signed on July 5, 1943, with work beginning the next month. With over 200,000 hours of work to build the ENIAC and a price tag of $500,000, the army didn't know what they were getting into. By 1944, the wiring diagrams for the ENIAC were completed, and construction could begin. To help with the wiring, Mauchly and Eckert hired female telephone operators. According to Goldstine (as quoted in *Proving Ground*): "Moonlighting telephone company workers, who didn't have much regular work, since the war was on and few companies were opening new offices, were hired to do much of the wiring inside the machine." While all 30 tons of the ENIAC were being built in the basement of the Moore School, Betty and the other computers could only guess what was happening, as they weren't allowed into the machine room. It was all part of a top-secret project, mysteriously titled "Project X," or "PRX."

That same year, in 1944, Betty was transferred to Pine Camp in Upstate New York, 15 miles from the Canadian border, to help with some computing work there. Pine Camp, now called Fort Drum, was created in 1908 and was home to the 10th Mountain Division of the U.S. military. While the frigid weather loomed outside, Betty worked on calculating weather balloon trajectories. According to Kleiman, Betty had no idea that these weather balloons were actually being used to deceive Axis forces. In fact, the U.S. military had many different projects happening at the same time to distract and trick the enemy. With intercepted messages leading to false trails or even entire fake troop units, the military hoped to reduce casualties while wearing down the enemy's morale. "Known as the 'Ghost Army,' these soldiers used inflatable tanks and artillery, sent false radio transmissions, and blasted audio recordings of troop movement and construction to create phantom forces that misled the enemy on American troop strength and locations," stated a 2020 article for the Defense Visual Information Distribution Service (DVIDS). This process was called "sonic deception," and it would only come to light fifty years later. The entire project was called a "traveling road show," as soldiers also had to learn how the weather changed the sound projections. They did this using several mobile weather stations. According to the book *Secret Soldiers,* the sonic deception proved to be extremely successful, as the fake troops could slowly replace real troops in an area without the enemy noticing, which saved thousands of lives. In the middle of these deceptions was Betty unknowingly working at Pine Camp.

Many of the fake audio recordings also came from Pine Camp, as the military worked with researchers at Bell Labs (see chapter 7) to create these tracks. The sounds were so real sounding that it was difficult for anyone to tell it was a fake recording. While Betty only stayed at Pine Camp for a few weeks, it would be decades before she finally learned what was actually happening there. For those in the Pine Camp area, many believed that the base was a series of "training grounds for armor units and the location of an internment camp for Italian and German prisoners of war," the DVIDS article added. "The post was generally believed to be inactive around 1944-1945 because the troops were all fighting overseas." Only fifty years later would the public, including Betty, come to know the truth of the projects at Pine Camp.

A Move to Aberdeen

When Betty returned to the Moore School, she was put in charge of teaching new recruits how to calculate trajectories. Meanwhile, construction had finally finished on the ENIAC, which included over 18,000 vacuum tubes, 10,000 capacitors, 1,500 relays, and 6,000 switches. To test the new computer, the army needed special programmers to test the equipment and learn to use it. As Project X was still highly top secret, Goldstine and the army decided to select five women Computers from the Moore School to learn the workings of the ENIAC. Another Computer would be added later, completing the "ENIAC 6." As Goldstine went around asking the best of the Moore School to join, it was no surprise that he asked Betty. As Betty recalled to Kleinman: "I happened to be a supervisor of the girls doing desk calculation for trajectories, and I got awfully tired of that, and so one night, Lieutenant Goldstine came and talked to us. He told us he couldn't tell us about the machine, because he said it was a secret project, but he asked us: would we be willing to go on this project?... I said to him, 'anything is better than what I'm now doing.' That's all I ever said. I never interviewed or anything." With that, Betty and her colleagues Kay, Jean, Marlyn, and Ruth became official computer programmers, with Fran Spence joining later. But before they could see the machine, the women were transferred down to Aberdeen Proving Ground, a Ballistics Research Laboratory base in Maryland. There they would spend six weeks training in the punch-card facility.

Aberdeen Proving Ground played a special role for the army as it was the place to test new equipment. As described in *Proving Ground,* with 27,000 soldiers living on the base, the five women didn't feel very welcome. Betty's friend and fellow Computer Jean explained to Kleiman: "Everywhere we went, we felt like a piece of meat surrounded by vultures." The women endured numerous catcalls and jeers. The dorms they slept in had no kitchen, so they often traveled to a local diner in Aberdeen's main square to enjoy hot meals. There, amid an unfriendly environment, the women grew closer. They talked about anything and everything in long, late-night discussions that Betty called "bull sessions." As the women had various religious and social backgrounds, they all learned more about different cultures from each other. Betty made these "bull sessions" more inviting with homemade cocktails. According to *Proving Ground*: "Betty had thought to bring her blender from home and she mixed frozen daiquiris that they drank as they talked together in one of their rooms, which they rotated."

Their days were full of learning plugboard components. During this time, according to Kleiman, Betty hounded one of the engineers to borrow the IBM manual to learn more about the ENIAC's plugboard. Since it was the Base's only copy, and the ENIAC was still top secret, the engineer could have lost his job if he let Betty borrow it, so he refused. That is, until one Friday when he cryptically mentioned to Betty that he would be leaving the manual on his desk during the weekend, and it had better be there when he returned. Taking the hint, Betty studied the manual the entire weekend, neatly returning it to its proper place before Monday morning. This was just one example to show how unique Betty's thought process was. As she told Ross: "I didn't know the first thing about electronics and I still to this day do not know electronics. I could follow a circuit but I couldn't tell you why it behaves [as] it does other than from the logic."

Betty wasn't the only one working hard to learn the unfamiliar terrain of the ENIAC's plugboard. According to fellow computer Jean Bartik in *ENIAC in Action:* "We spent much of our time at APG [Aberdeen Proving Ground] learning how to wire the control boards for various punch card machines: tabulator, sorter, reader, reproducer, and punch." This group learning would be helpful in learning to work the ENIAC.

Testing the ENIAC

While the women were in Maryland, the ENIAC was still being built and tested. The vacuum tubes were particularly tricky to implement, as they often blew. "Like the light bulbs from which they had evolved, vacuum tubes would burn out after an unpredictable amount of time, usually as they were turned on," explained *ENIAC in Action.* Because of this weakness, the ENIAC engineers quickly developed a system to pick the strongest tubes for the machine. "Tubes weren't all created equal, and the weak ones usually failed early in their lives," according to *ENIAC in Action.* "Eckert's team built special equipment to test incoming tubes, verifying the quality of each batch and accumulating data based on their failure patterns. The results were fed back into the design process." The problems with the tubes were also related to how much current was running through them, so Eckert ordered that no more than 25% of the recommended current amount (by the manufacturer) be run through the tubes. The tubes also blew most often as the power was turned on and they were warming up. For this reason, "ENIAC was to be turned off as little as possible; the power was to be left on overnight and on weekends," stated *ENIAC in Action.*

Upon returning to the Moore School, the ENIAC 6 received some big news. Mauchly, Holberton, Eckert, and the other ENIAC engineers were finally able to show the women the true identity of Project X. It was a wonder to behold. With a U-shaped layout running 80 feet long, the women could see all 45 units of the machine. 16 units ran on each side with eight at the bottom of the U-shape. The other five units were mobile and could be moved around on carts. It could also be accessed from the front and the back, which would be helpful for programming. It was a behemoth, and the women were about to learn how to use it.

Because the ENIAC had no internal storage, each new calculation had to be programmed into the machine by hand. Four subunits made up the computer: an accumulator (which did the functions such as adding or dividing); the function tables (which gave the ENIAC a read-only memory); the governing units which controlled the other units (like a "cycling unit" which started operations for other units); and the units of input and output, punch cards. According to a 1996 article in the *Wall Street Journal*: "Running the ENIAC required setting dozens of dials and plugging a ganglia of heavy black cables into the face of the machine, a different configuration for every problem... Every datum and instruction had to reach a correct location in time for the operation that depended on it, within 1/5,000 of a second." To use the cables and switches, the women looked at programming maps that told them which switches and cables to

move for specific calculations. According to the *Computer History* film, "Mapping the [computer] programs out on paper took weeks." To then set the switches and cables to match the "program diagram took many days," the film also stated.
The other challenge with the ENIAC is that it ran parallel programming, where multiple programs could happen at the same time. This required multiple switches and wires to be plugged in quick succession, as the machine would break if things went too slowly.

With these complex units and the need for coordinated and rapid wire movements, the ENIAC 6 had their work cut out for them. "We weren't allowed to write anything," Betty told Ross, in describing the process of learning to program the ENIAC. "We were not allowed to have anything; we were not allowed to keep notes or anything at all." This made learning ENIAC programming difficult, especially for Betty. Thankfully her logical thinking processes helped her and her colleagues to break down the programming process into distinct parts, which would in turn be useful in later creating specific programs for the ENIAC.

Betty's logical thinking became especially important when fixing the many problems that happened during the ENIAC's programming, such as the vacuum tubes breaking. The women got so good at running the computer that they could eventually pinpoint which specific vacuum tube out of the 18,000 had broken, as Betty told Ross. The programming itself would also break mid-run, something that coders still deal with today. Betty came up with a logical method to narrow down the problem and find the solution, where they would run the code and examine where in the wires it had broken.

The ENIAC Comes to Life

The first big project for the ENIAC came as a request from Los Alamos National Laboratory. As the Manhattan Project was in full swing, many scientists were beginning to investigate atomic bombs as potential weapons. One of these scientists, Edward Teller, was interested in using tritium, an isotope of hydrogen, as fuel for the bomb. As tritium could "ignite" in a self-sustaining chain reaction, it could be a perfect source for the bomb. However, there was still an important question to answer: "what was the minimum amount of tritium needed to produce an ignitable mix," *ENIAC in Action* stated. An actual experiment was not possible, as creating tritium was an expensive and long process. Instead, Teller hoped that the ENIAC could simulate an experiment by running several calculations needed to answer this question. As the machine was too big to move to Los Alamos, Teller sent two scientists to the Moore School. "One interesting feature of early ENIAC practice is the machine's ability, until about 1952, to attract what participants called 'expeditions' or 'pilgrimages' by scientists eager to apply its unique talents to their own problems," explained *ENIAC in Action.* The Los Alamos project did just that, as Betty and her fellow programmers helped to solve several calculations for the two scientists, without being told what the project was about. Supposedly a million punch cards were sent back to New Mexico with the two scientists when the project was completed. While this number may seem like a lot, the ENIAC required large amounts of punch cards for both inputs and outputs, as well as temporary storage units.

After the Los Alamos project, the ENIAC 6 team were then tasked to create a program to calculate ballistics trajectories, just as they were doing before. Betty worked closely with Jean to develop the logic for the program. To do that they made special charts called "pedaling sheets" to help work through problems. "[Jean] speculated that the operators called these 'pedaling sheets' (perhaps…because they showed us what was happening at each "pedaling stroke" along the way, or add time by add time),'" *ENIAC in Action* stated. "This suggests a pun on the name of ENIAC's cycling unit, which provided the ability to step through a program for debugging purposes." These pedaling sheets worked like a flow chart and enabled faster movement for the programmers. Betty later told Kleiman that programming the ENIAC was "probably one of the hardest things I ever did." An initial demonstration was given to the army and others on February 1946. Betty and her colleagues, called the "operating girls" according to a planning document, were asked to give tours, answer any questions, and introduce the engineers. They were not asked to explain their own work on the project. The women also helped serve lunch and no one realized that they were the real minds behind getting the machine to work. All the invited journalists and guests left with an ENIAC punch card as a souvenir. After a successful initial demonstration, the women were asked to then demonstrate their ballistic trajectory program for an audience. On Valentine's Day, the women worked all day to finalize the program before demonstration day.

On the morning of February 15, demonstration day, Betty arrived at the Moore School excited. She finished tweaking the dials and announced that the program was complete, to the amazement of her cohort. Betty explained that in her sleep, she had run through the entire program and narrowed down on which wires to fix. As Jean later told Kleiman, Betty did "more programming in her sleep than people did awake." The demonstration began promptly at 11 am and several of ENIAC's male engineers ran the machine while the women sat at the back of the room. Everyone cheered after the calculations came out correct, but no one congratulated the women. Betty felt resentment for the entire setup of the demonstration, stating to Kleiman that it was a "man's show... We expected that. In those days women were not recognized at all. So, it was just a normal thing."

With the successful demonstration, the top-secret part of the project was officially over, as the world now heard about its first modern computer. Betty was finally able to tell her family and given their scientific interests, they were very proud of her. The media called the ENIAC a "giant brain." Because of its mysterious nature, many early myths about the ENIAC arose during this time, including one about all the lights in Philadelphia dimming when the giant machine was flipped on. These myths in turn inspired many science fiction authors who "perpetuated that image, often assuming that increasing computing power demanded increasing size," *ENIAC in Action* stated. "Isaac Asimov wrote stories about the Multivac, a computer with people and rooms inside it that grew to fill an entire city and run the world." Clearly, there were many different opinions about what this amazing machine could do.

After the demonstration, things became a bit tumultuous, as Mauchly and Eckert got into a patent dispute with the Moore School and the army. As *Computer History* explained, the University of Pennsylvania had adopted a new patent policy in 1946, which would have forced Mauchly and Eckert to turn over all their patents on the ENIAC to the University if they stayed past March. These patent disputes tore a rift in the original ENIAC team, as the women programmers sided with Mauchly and Eckert while Goldstine and others argued against them. These disputes eroded relationships between these brilliant individuals, and worse, created negative publicity for the entire project. According to *ENIAC in Action:* "From the 1950s to the 1970s, lawyers did far more than historians to keep ENIAC in the public eye. The narratives they constructed for an audience of judges and patent officials would determine the fate of millions of dollars in licensing fees." The fees weren't the only important topic of debate in these disputes, as both sides argued that they were influential in "inventing" the first computer. Ultimately these disputes were never fully resolved.

Because of this both Mauchly and Eckert left in March 1946 to start their own company but returned in June of 1946 to present the Moore School lectures, a series of lectures on the ENIAC. Historians credit these lectures with being influential in spreading the "stored program concept" for future computers. These lectures were attended by some of the smartest researchers in the U.S., including mathematician Claude Shannon of Bell Labs (see chapter 6). While the lectures were closed to women, Betty listened through an open window of an adjoining room as history was being taught, according to *Proving Ground.* Who knows, Betty and Claude could have met, though there is no record if they did.

In late 1946, due to its hefty price tag, the army paid for and took control of the ENIAC, deciding to move it to the Aberdeen Proving Ground Base in Maryland, 75 miles from the Moore School. The Moore School was awarded $96,200 ($1.4 million today) to relocate the machine to its new home. This cost would go towards fixing the ENIAC after a small fire had damaged one of its panels, as well as the entire move and installation process. The computer was temporarily shut down on November 9, 1946, for refurbishment and transportation. The entire trip would take several months, and the ENIAC would be turned back on in Aberdeen on July 29, 1947. During its move, Betty and the other programmers had several months off. During this time, Betty, Kay, and Holberton decided to take a long road trip around the country, according to Kleiman. They saw the Pacific Ocean, Yosemite National Park, and the central U.S., and even visited Holberton's family in Virginia. When the trio returned, it was just in time to move to Aberdeen.

While during the week Betty taught ENIAC programming at Aberdeen, on the weekends she helped Eckert and Mauchly build the UNIVAC (standing for Universal Automatic Computer). To build and sell the UNIVACs, Eckart and Muchly created the Eckert Mauchly Computing Company (EMCC or as Kenna read it, E= MC^2). In late 1947, the company finally had enough money to hire Betty and she joined EMCC as employee #13. She said tearful goodbyes to the rest of the ENIAC 6, but by that point, many of them were leaving Aberdeen as well. With her background in programming the ENIAC, combined with Eckert's and Mauchly's engineering skills, the team made some significant changes to the UNIVAC to make it easier to use, especially to its memory system. "Our whole philosophy was we went from twenty accumulators on the ENIAC to a thousand memory locations in the UNIVAC," Betty told Ross. "That was a big jump...we learned how to manipulate that external memory very well."

One of Betty's biggest contributions to the UNIVAC, and computer science in general, was developing the C-10 programming code for the UNIVAC. As Betty was passionate about making computing accessible to everyone, she made the C-10 code mnemonic, so commands like "a" meant "add" while "b" meant "bring." Computers still use a similar process (e.g. control + p= "print"). According to the 2001 *Chicago Tribune* obituary about Betty, she made C-10 unique by using "keyboard commands rather than by dials and switches." The article continued by explaining that "she designed control panels that put the numeric keypad next to the keyboard and persuaded the engineers to replace the UNIVAC's black exterior with a gray beige tone that is now the universal color of computers." By implementing these changes, Betty was trying to make sure that these computers could be accessible to almost anyone wanting to learn. Betty also focused on creating new software that would make computing a more efficient process. According to a 2000 NIST special publication, Betty "also wrote the first sort-merge software, notable as perhaps the first case of a program whose function was to generate another program..." As the first UNIVAC models were sold to Prudential Insurance, Nielsen (a national survey company), and the U.S. Census Bureau, Betty made sure her C-10 coding instructions went with each model; to make it easier for others to use.

1950 became a big year for Betty. In July, she married John Vaughn Holberton in Philadelphia. Priscilla, Betty's daughter, told Kenna that "the wedding photographer either never showed up or they never got their photos. I can't remember exactly." The only photos from Betty's wedding day are blurry ones taken by her friends. After their marriage, Betty would go on to give birth to two daughters: Pamela and Priscilla. Priscilla remembered Betty working hard to become an involved mother. "She was the room mother [a parent assistant for the teacher] in my sister's first-grade classroom," Priscilla told Kenna. "Pamela had a black first-grade teacher and no other moms volunteered to be room mothers. I didn't know this at the time." Phil Holberton, Betty's nephew, recalled that Betty and John seemed to be somewhat opposite personalities, as John was the extrovert and Betty the introvert. "Even though Uncle Vaughan [as Phil knew him] was very deaf, he had a twinkle in his eye and enthusiastically engaged with others," Phil recalled. "An amiable, kind man who was profoundly deaf but used a microphone and headset to hear others. He worked very hard at overcoming deafness and not being isolated. They both were very bright, but Betty was the cerebral one. She had a very analytical approach to life; for instance, she tackled her family's genealogy as a hobby." These opposites seemed to complement each other as Betty and John remained happily together until Betty's death on December 8, 2001. John would, unfortunately, pass away a couple of years after this on April 15, 2003.

1950 also brought a significant work change for Betty, as EMCC ran out of money. Though they had three orders already for UNIVAC models, they had underestimated the price, merely using the number of vacuum tubes as the cost of the model, as Betty explained to Ross. The company Remington-Rand Inc. bought out EMCC and continued selling UNIVAC models. In 1952, the UNIVAC once again made headlines when it correctly predicted that General Dwight D. Eisenhower would win the presidential election. The company panicked, worried that they had swayed voters, yet the win for Eisenhower was still nationally accepted.

At EMCC and then Remington-Rand, Betty worked alongside other talented female programmers including Grace Hopper. Hopper is perhaps one of the biggest names for women in computer science, as she helped develop the computing language COBOL. In a 1967 *Cosmopolitan* magazine article, Hopper is quoted as saying that programming was "just like planning a dinner. Programming requires patience and the ability to handle detail. Women are 'naturals' at computer programing." Like Betty, Hopper made sure to mentor and support other women at both EMCC and Remington-Rand. One female employee, Adele Mildred Koss recollected in a 2003 article that Hopper had brought in water guns and an umbrella to work one day to cool off the workers sitting in the hot machine room. When Koss got pregnant, Hopper told her to "take her work home with her," allowing her to continue working while managing a family. This was unheard of for a woman in the 1950s but showed that EMCC made sure to offer a flexible and fun workplace for its large number of female workers. Both Hopper and Betty worked on the programming, leadership, and strategy for EMCC, overseeing their own teams of talented programmers, which was a remarkable occurrence to have these two amazing female computers working at one company.

The flexible and supportive company culture changed when Remington-Rand took over. Suddenly it felt like women were no longer respected equals. As Betty told Ross: "The salespeople got in and the whole thing changed. Women, as far as I could see, had absolutely no place under Remington-Rand, absolutely none... This was so annoying that I could see the writing on the wall..." With no Ph.D. and a journalism degree, Betty worried about her future career. She didn't want to stay in a company where saleswomen were seen as glorified secretaries, but she wasn't sure what to do. She finally saw an opportunity when the US Navy's David Taylor Model Basin, a testing facility, was having trouble understanding how to program the UNIVAC. She began working there, first as a consultant, and later as a full-fledged employee in charge of writing instructional codes.

From 1953 to 1966, Betty was a member of the Applied Mathematics Lab at David Taylor. There she made sure the user had all the instructions they might need to run the UNIVAC. She then spent around two to three years tackling the UNIVAC LARC (Livermore Advanced Research Computer), an early supercomputer. With LARC, Betty helped design and test its software, making it more accessible to scientists. Betty also did many more amazing things at this facility, which is hard to account for in just one book chapter about her life.

In 1966, Betty transitioned to the NBS to work on regulating and standardizing computer code. By the mid-1960s coding languages like Fortran and COBOL were becoming more widely used, yet they weren't standardized, leading to software that didn't work reliably. Betty and her team created a series of test programs to test if the code worked, establishing a set of standards for how coding should operate. As the world of computers became more mainstream, Betty saw that there might be issues, "I felt that human communication was our basic problem," she told Ross. "Whenever we went anywhere, people couldn't understand what we were saying; the vocabulary was bizarre (it still is). We were always making up terms... very much a parochial kind of attitude in the computer field and I think very much today." Thanks to her insight into how users may see these new computing languages, she was able to make them more accessible to anyone interested in learning about computing, helping to usher in the new era of digital computers.

In 1983 Betty officially retired. There was little fanfare as few people knew the full extent of her work. It wasn't until 1996 that Kleiman, a lawyer, and a computer scientist herself, found out the story of the ENIAC 6 and went to work publicizing it. "They were shocked to be discovered," Kleiman said in a 2020 article. "They were thrilled to be recognized but had mixed impressions about how they felt about being ignored for so long." Kleiman made sure that the ENIAC 6 were at the 50th anniversary of the ENIAC in 1996. Kleiman was not the only one wanting to share these women's stories, as "a special film featuring the women, LeAnn Erikson's *Top Secret Rosies* premiered in 2010," according to *ENIAC in Action.* Kleiman released her own film about the ENIAC 6 at the 2014 Seattle film festival.

Now with her story made public, Betty's achievements could be recognized. In 1997, she and the rest of the ENIAC 6 were inducted into the Women in Technology Institute's (WITI) hall of fame. That same year the Association for Women in Computing presented Betty with the Augusta Ada Lovelace Award. Also in 1997, Betty also received the IEEE Computer Pioneer Award for "her developing the sort-merge generator which IEEE told its members 'inspired the first ideas about compilation.'"

On Dec 8, 2001, Betty passed away after recovering from a stroke. She continued to share her wisdom and her story until the day she died. From programming the first modern computer to standardizing coding languages, Betty's 40 years of work in computing make her a legend. Her kind heart and caring spirit also helped support her colleagues and subordinates in ways no one else could. Betty's story not only illustrates her qualities as a role model and leader but also reveals the skills that many young scientists need in order to succeed, such as determination and logical thinking. For these reasons and many more, Betty's story, along with the rest of the ENIAC 6, should be embraced in the public's consciousness.

Elisa Torres Durney, the leader of "Girls in Quantum," a global non-profit organization, shared with Kenna how Betty specifically has impacted her own life. "Women's participation in computing is critical not only to close the devastating gender gap in STEM but also to empowering more women and girls," 17-year-old Durney stated. "For me, it means having role models such as Betty Holberton they can look up to and be inspired by. There is no set gender for a career; instead, we should encourage women to pursue their interests by providing them with the necessary tools, opportunities, and support." Durney is just one of many women, particularly young women, who have found

Betty and her cohort of women programmers to be inspiring role models.

Both Brian and Kenna are proud to work with and showcase the work of many women leaders in the field of quantum science. In his own work at the Quantum Strategy Institute (QSI), where the majority of leaders are women, Brian encourages and supports participation in the quantum field from all walks of life and all geographies. With many countries and continents just beginning their quantum journey in earnest such as India, Africa and the Middle East, encouraging large and small efforts generates greater likelihood of long-term quantum success. For her part, Kenna writes an article series called "Women of Quantum Technology" for *Inside Quantum Technology,* which highlights many of the women working in both the quantum industry and in academia. Betty Holberton's example and those of her peers and successors drive excitement and greater confidence around the possibilities and options for women and girls worldwide.

Lorraine Tsitsi Majiri, the President of OneQuantum Africa, an African-based quantum computing organization, also highlighted the importance of Betty's role in the technology industry. "Women have made significant contributions to the tech industry and continue to play a vital role in its growth and success," Lorraine told Kenna. "Despite being underrepresented in the industry, women have played key roles in the development of many major technological innovations, such as the first programming language, the first computer game, and the first spreadsheet software. In recent years, there has been an increased focus on promoting diversity and inclusion in the tech industry, and many women have taken on leadership roles in major tech companies and startups. For example, Meg Whitman is the CEO of Quibi, a new streaming service focused on short-form content, and Ilana Wisby is the CEO of Oxford Quantum Circuits.

Overall, the contributions of women to the tech industry are significant and their presence and leadership are essential to the industry's continued growth and success. Knowing that other women have been and continue to be successful in the quantum technologies sector inspires me to not give up on my own goals in quantum computing. These ladies have helped me, and it has inspired me to pay it forward."

Chapter 8

Pantur Silaban - "Mr. Relativity"

The lecture hall is filled as students, scientists, and the general public all settle in. Before them stands Pantur Silaban, a thin man in a long sleeve shirt and long pants. With a piece of chalk in one hand and a smoking cigarette in the other, Pantur begins writing on the chalkboard. A silence hangs over the audience as they read in big letters: "The Age of the Universe." Pantur turns to them, smiling.

"Let us begin," he says.

***

Out of the ten "giants" we focus on in this book, Pantur Silaban was possibly the hardest to write about. Not only is the type of physics he studied extremely difficult to understand and convey, but there are only a handful of sources about Pantur, and most are in Indonesian, thus requiring translation.

Pantur's story ends rather recently, as he, unfortunately, passed away on August 1, 2022, at the age of 84. With his research focused on trying to unify quantum theory with general relativity, Pantur was a leader to both the Indonesian scientific community, and the physics community as a whole. He was an inspiring mentor to his many students and developed a reputation for being a jokester. Transitioning from the traditional society of Indonesia to the fast-paced campus of the University of Syracuse in New York, Pantur navigated many different challenges in his life. While his name may not be recognizable to many of us, Pantur's work is continuing to be advanced by quantum physicists all over the world. Because of his work, a couple of writers crowned him "Mr. Relativity."

Author's note: While neither Brian nor Kenna have the technical background in the theoretical physics environs Pantur focused on, we will do our best to explain the importance of his work. All direct quotes from Pantur and his colleagues were originally in Indonesian and have been translated.

Pantur was born on November 11, 1937, in Sumatra, Indonesia. Unlike most of the "giants" covered in this book, Pantur's parents, Israel Silaban and Regina Lumbantoruan were illiterate yet inspired Silaban towards his eventual career in their own ways. Brian found Pantur's story especially relatable given his own parents had only a grade eight education back in the early 1940's, yet they still inspired Brian to reach for his goals in executive roles in banking, leadership and advanced technology.

While there is little information on Pantur's early life, we do know that he became interested in physics and decided to get his bachelor's degree in the subject from the Bandung Institute of Technology (ITB), a well-known university in the area. This would have no doubt been a huge change for his parents, who probably did not understand what going to a university would entail. However, we do know that Pantur's parents were very supportive of his choice. According to a 2004 interview, Pantur explained that "my father, who was a trader and illiterate, only said 'it's up to you to choose. We can only help send you to school. My advice is to choose a field that you like.' My mother, who is illiterate but good at counting money, suggested that I choose medicine. Apparently, she saw that every time a doctor treats a patient, the doctor gets the money." With the support of his parents, Pantur enrolled at ITB to begin studying physics.

Thanks to this support from his parents and his own diligence, Pantur successfully graduated from ITB in 1964 with a B.S. in physics. After being exposed to many types of sub-fields of physics, Pantur settled on theoretical quantum physics as his main interest. To study this field further, he knew he would have to leave Indonesia and travel elsewhere, as Indonesia did not have the educational resources to teach him (for context Pantur would become the first Ph.D. graduate in Indonesia to study theoretical quantum physics). Looking at potential universities to continue his education, Pantur settled on the University of Syracuse in New York. Pantur knew Syracuse would be a perfect choice as it was the first university in the U.S. to have a research group solely focused on Einstein's theory of relativity. Having made his decision, Pantur packed his bags, said goodbye to his parents and community, and set off for New York.

A Dive Into General Relativity

Albert Einstein's famous theory of general relativity was developed between 1907 and 1915, decades before Pantur would study it. While the theory posits many different ideas, it mainly deals with gravity. Einstein suggested that space-time was similar to a fabric, which meant that it could be warped by mass, a process otherwise known as gravity. While it's a simplification, it might help to think of a watermelon and a grapefruit sitting in a suspended hammock. Both fruits "warp" the hammock's fabric, but the warping is greater with the watermelon due to the watermelon's heavier mass. As researcher Andrew Zimmerman Jones explained in a 2020 article: "The curvature of a light object doesn't affect the heavy object as much, but the curvature created by a heavy object is what keeps us from floating into space. The curvature created by the Earth keeps the Moon in orbit, but at the same time, the curvature created by the Moon is enough to affect the tides." With this illustration, Einstein originally posited that the force of gravity was due to the curvature of the space-time fabric itself.

Since 1915, when the theory was published, scientists had been working tirelessly to both validate Einstein's theory of relativity experimentally and to also use it to predict other processes. One of these scientists was Peter Bergmann, who was Einstein's assistant for a period of time and would become Pantur's Ph.D. advisor at the University of Syracuse. Born into a Jewish family in Germany, Bergmann was trying to finish his Ph.D. when the Nazi regime came to power. "Concerned that Peter would not be able to continue his studies in Nazi Germany, his mother secretly wrote to Albert Einstein in the summer of 1933 to ask where he would consider accepting her son as a Ph.D. candidate," explained a 2003 obituary for Bergmann in *Physics Today*. "Einstein demurred and suggested that Peter study with Wolfgang Pauli first." Even with Einstein's refusal,

Bergmann achieved his Ph.D. at age 21 from the German University in Prague. He then wrote to Einstein directly, not realizing that his mother had already done so, and asked to be his assistant. Einstein agreed and in 1935 Bergmann fled Nazi Germany for New York. The rest of his family was scattered around Europe and his sister Clara would later die at the Auschwitz concentration camp.

From 1936 to 1941, Bergmann worked under Einstein at the Institute of Advanced Study (see chapter 5). He then became a professor at the University of Syracuse from 1946 to 1981, where he would advise 32 Ph.D. students, including Pantur. In 1942, at age 27, Bergmann published the *Introduction to the Theory of Relativity*, with a foreword by Einstein. In the foreword, Einstein wrote: "Much effort has gone into making this book logically and pedagogically satisfactory, and Dr. Bergman has spent many hours with me which were devoted to this end." Because of Bergman's close relationship with Einstein, Einstein also was closely connected to the University of Syracuse and kept an eye on the goings-on there. As a Syracuse professor, Bergmann not only reflected the type of mentorship that Einstein had given him but also went the extra mile in guiding his students. According to *Physics Today*: "His sensitivity to students and their need for support is demonstrated by the following anecdote. Once, during a faculty meeting, Peter was asked how he came to give an A grade to a student who was failing most of his other courses. Peter left the meeting, examined his grade book, returned, and responded, 'It was a low A.'" With his considerate and supportive attitude toward his students, Bergmann easily made an impression on all of them, including Pantur.

Bergmann was also responsible for making the University of Syracuse's Physics Department one of the best in the world for studying theoretical physics. Using his extensive connections, Bergmann was able to create the first research center in the U.S. focused on general relativity. Within the center, Bergmann focused on trying to reconcile general relativity with quantum theories, into what experts call the "Grand Unified Theory" (GUT). This would be what Pantur would focus on when he arrived in New York,

Unifying the Theories

In 1967, Pantur began studying under Peter Bergmann as well as Syracuse professor Joshua Goldberg. All three were interested in trying to connect quantum theory with Einstein's theory of relativity into General Unified Theory (GUT). In order to explain Pantur's work, we will have to dig into what GUT actually is and why it matters. Though neither Kenna nor Brian is a theoretical physicist, we have done extensive research on these theories for this chapter and will do our best to explain them in an engaging way.

In physics, the Standard Model describes the four main forces that interact in nature: the long-range weak force, gravity, the strong force, and electromagnetism. The weak force (or weak nuclear force) is responsible for particle decay, such as in radioactive isotopes, while gravity is considered the attraction between two objects. The strong force (or strong nuclear force) binds fundamental particles together, such as keeping protons and neutrons to ether in the nucleus. Finally, the electromagnetic force works on charged particles, like protons and electrons, causing them to attract and repel. While these four forces are distinct from each other, physicists have theorized that the electromagnetic and weak forces can actually be unified at a high energy scale into an "electroweak" force. This meant that the mathematics of each force was essentially the same, and both were constrained by the same issues. However, at low energy, the two forces could appear separate and measurable.

In 1974, physicists Sheldon Glashow and Howard Georgi extended the theory to also include the strong force. They "discovered that all of the known particles and three non-gravitational forces could naturally fit within a single symmetry structure," explained a 2017 *Nautilus* magazine article. They called this structure the Grand Unification Theory or GUT. From their calculations, Glashow and Georgi also wondered, since the forces separated at smaller scales if there could be a scale small enough where all three separate forces had identical interaction strengths. Doing further research, they suggested that this was possible, but only at a level 15 orders of magnitude smaller than the size of the proton. With this constraint, the physicists realized it would take a giant accelerator, approximately the size of the Moon, to experimentally test this.

Building off of Glashow's and Georgi's ideas, other physicists suggested a different experimental setup looking at proton decay (where a proton breaks down into smaller subatomic particles). According to the *Nautilus* article: "If the Grand Unified Theory predicted a proton lifetime of say, a thousand billion billion billion years, then, if one could put a thousand billion billion billion protons in a single detector, on average one of them would decay each year." This setup made both the timescale and the energy scale much more measurable. To test this, scientists set up experiments at Lake Erie and in a mine near Kamioka, Japan in 1982, using 3,000 tons of water to supply the protons. "So, all that was required was to get a tank of water, put it in the dark, make sure there were no radioactivity backgrounds, surround it with sensitive phototubes that can detect flashes of light in the detector, and then wait for a year to see if a burst of light happened when a proton decayed," the *Nautilus* article added. Because one proton was supposed to decay every year, the experiment ran for several years. In the end, the results were...inconclusive. Instead of a proton decaying, a different subatomic particle called a neutrino decayed, which the Standard Model didn't account for, so the results didn't prove or disprove the GUT theory (though they did help advance neutrino physics).

There are still many ongoing experiments to try and validate the GUT, but it might take a while. Experts believe that this unification is probably possible but "some new physics at energy scales beyond those that have been observed thus far must exist," the *Nautilus* article stated. In short, it'll take many more physics advancements before we can see if GUT can be validated.

Resolving Quantum Gravity

The work done by Glashow and Georgi as well as others was part of a trend in theory unification. In the late 1960s to early 1970s, Pantur was at the beginning of the trend. While physicists were looking at unifying the three non-gravitational forces,

Pantur decided to focus on gravity, specifically in the context of relativity. As Pantur explained in 2004:

> "In nature, we know there are four kinds of interactions: gravitational, electromagnetic, strong, and weak. Gravity is the longest-known phenomenon, but even now it is still a mystery. The first theory of gravity stems from the ancient Greek view that an object falls to Earth because it belongs to Earth. [This is] similar to the expression we know. If you own it, you won't go anywhere."

As we discussed earlier, the theory of general relativity describes gravity as the warping of the space-time fabric (think fruits in a hammock). In contrast, the quantum theory (or quantum field theory) describes gravity as a particle. According to a 2022 article from JRank Science & Philosophy: "quantum theory depicts a quantum field with a carrier particle for the gravitational force that—although not yet discovered—is termed a graviton. In stark contrast, general relativity theory does away with the need for the graviton by depicting gravity as a consequence of the warping or bending of space-time matter (or more specifically, mass)." In order to reconcile these theories physicists, like Pantur, have to somehow combine a gravity particle and a gravity warp. As you can imagine, this is still being heavily studied, and like the GUT experiments, may be limited by our current technology.

Returning to Indonesia

After completing his Ph.D. in 1971, Pantur returned to Indonesia and became a professor of physics at ITB. Like Bergmann, he made sure to support his students as best as possible, especially since the science they were working on was extremely abstract. "Learning physics, in particular, there is the phrase: 'Do not cover it, discover it,'" Pantur said. "I always try to enter undergraduate and graduate classes without a piece of paper, even if it's about the quantum field and general relativity." As a professor, Pantur emphasized the importance of discussions in his classes, and his students asking questions instead of him just lecturing at them. He made an intriguing figure, as he was described as smoking, wearing a "long-sleeved shirt and trousers...that never wrinkle," all while scribbling notes on a chalkboard. Pantur also gained a reputation as a jokester, telling many jokes throughout his lessons. "I like to hear and read jokes that have a moral to the story," he said. "Sometimes I create them myself."

Sometime while at ITB, Pantur married Rugon Lumbantoruan (who may have been related to his mother, since they share a last name). He had four daughters with her, all of whom graduated from local universities. Through funding his daughters' studies, Pantur showed his support for education, especially for those who have been traditionally excluded from education opportunities due to religious reasons. One of his daughters had a grandson, whom Pantur named Joshua after his old professor Joshua Goldberg.

Throughout the rest of his life, Pantur continued to work uniting general relativity and quantum physics. In 2004, he was asked by ITB's Chancellor to give a public lecture titled the "Age of the Universe," which over 300 people attended. Various physicists, such as Roger Penrose, have cited Pantur's work. Pantur was also involved in physics organizations, both domestically and internationally. This included the "Indonesian Physical Society (HFI); South East ASEAN Theoretical Physics Association, Singapore; Society for Gravitation, Boston, Massachusetts, USA; American Physical Society (APS), USA (until 1987); and the Albert Einstein Society, Bern Switzerland," explained a 2022 article from ITB.

While Pantur retired in 2000, he continued to do research until the day he died. According to his colleague and ITB professor Dr. Freddy Permana Zen, Pantur still went to his office at ITB and "consistently did research in the field of theoretical physics even though there is no funding for that research." On August 1, 2022, Pantur peacefully passed away at the age of 84, in the Santosa Hospital in Bandung. Many at ITB mourned his passing. As the Deputy Chancellor for Finance, Planning, and Development, Dr. Ir. Muhammad Abduls stated: "Prof. Pantur Silaban, Ph.D., was an ITB lecturer whose various works we need to refer to and whose attitude in life needs to be emulated as an example."

Pantur's legacy lives on in the research that other scientists work on. As many physicists are designing quantum sensors to detect Earth's gravity (also called gravitometers), they use Pantur's work to help develop these devices. Others are using quantum theories to help advance quantum computers as well as different electronic machines. Even at ITB, Zen and his team are looking at quantum simulations and how environmental noise affects those systems to help improve many different devices. As Zen stated in a 2022 article for ITB. "Every electronic device that we are using right now is made from materials that were invented through the application of quantum theory."

Pantur's story also helped shape the entire debate on the Grand Unified Theory. Before Pantur, scientists with Nobels (and mainly from western countries) were the overwhelming majority of those looking at this field. "Yet, nearly a half-century later Prof. Pantur Silaban injected himself from the most unlikely situation (an Indonesian, surely disregarded by others in this field) and was able to rise above his perceived lack of blue blood pedigree, and made several significant discoveries..." explained Jason Razien, the founder and General Manager for Clockworks Technologies, which focuses on cybersecurity. Pantur not only helped to advance many physical theories but also helped to pave the way for many more physicists from different backgrounds to enter the field. Pantur believed in this idea of widening accessibility and even highlighted the importance of listening to everyone. He also emphasized that traditional role models were somewhat outdated. As he put it in 2004: "In my opinion, no matter how respected a person is, there is much that he does not know, but is known by people who are completely uneducated. I want to break the tradition of role models."

Chapter 9

Mark A. Reed – "The Quantum Cowboy"

The hotel's fluorescent lights flicker in a distracting manner as Dr. Mark Reed settles into a leather sofa in the lobby. The cliché Bossanova music is doing little to calm him, as he taps his foot anxiously against the garish orange shag carpet and continues to look at the clock.

"Do you have all your notes?" a voice asks. Mark turns to see a senior colleague, coffee-in-hand, peering down at him.

"Yes, though, I'm waiting to hear from Dallas this morning."

"Oh?" The colleague asks, settling across from him. "I thought you had your results already. I mean the conference is in only two hours, it's a bit late to change anything."

"I know but we're trying one last test, to try to nail down the right size for our dots," Mark responds. "We hope to…" he trails off as the phone on the concierge desk begins to ring. A clerk picks it up and after a moment, locks eyes with Mark. She waves him over and he rushes to the receiver. His colleague watches as a few moments of silence pass before a grin breaks across Mark's face. He hangs up the phone and dashes to grab his briefcase.

"I'm sorry, I have to leave."

"Good news?" The colleague asks.

"Yes! Which means I have to go rewrite a few things," Mark says as he rushes for the hotel room that reads "fax machines."

***

While quantum computing is the main focus of this book, it's important to note that quantum computing is just one of many types of quantum technology. As quantum mechanics govern every process, there have been scientists and researchers over the years who have leveraged these processes to create innovative devices. From quantum batteries, which promise more efficient energy storage (with large implications for things like electric vehicles); to atomic clocks, the most accurate clocks on the planet that keep our society on time, these other devices are just as important as quantum computers in bettering society. As such, Kenna and Brian thought it only fair to include one of the key innovators for these devices in this book: Dr. Mark A Reed, who coined the term "quantum dots," and helped to create the first type of "microfabricated" quantum dot, forever revolutionizing nanotechnology. Not only did Mark's work make strides in what nanotechnology can do, but his passion for supporting and inspiring others, especially students, has helped to transform many careers.

As Mark passed away on May 5, 2021, parts of his story are still unfolding. His colleagues at Yale University have published a virtual tribute wall for those who have been impacted by Mark to post their stories. We will include a few of these tributes at the end of the chapter. With Mark's death still affecting many lives, we hope to do his story justice within this chapter and illustrate the example Mark posed as a scientist and mentor. Kenna thanks the students and colleagues who have shared their stories about Mark with her, showing just how much he will be missed.

Born on January 4, 1955, in Suffern, New York, Mark began his life by enjoying the nature around him. While there isn't much public information about Mark, we do know he was very active. His wife, Elizabeth J. Reed, stated in an obituary for *Nature Nanotechnology*: "Outside of school, he enjoyed hunting, and fishing in upstate New York and learned woodworking from his father. The black walnut furniture in his home was designed by Mark and his father, the wood originated from trees he felled with his father and brother." Mark studied originally at Christian Brothers Academy, then went to Syracuse University for his Bachelor of Arts degree, Master's degree, and Ph.D., all in physics. In high school, he stayed particularly busy. "Throughout his school years, he belonged to chess teams, played football, played the piano, and flute, and sang in a student band," explained an obituary from Yale. Several of these hobbies Mark continued later in life, like chess, where he would study professional matches to improve his own skills. Mark was also interested in astronomy. According to Elizabeth Reed's report for the *Nature Nanotechnology* obituary: "No matter where in the world he was, he could name the constellations, know when there would be celestial events such as meteor showers, eclipses, space station sighting. Day or night you could find him patiently waiting, staring at the stars to catch a glimpse of the predicted astronomical event. One of the trips to Hawaii included a memorable private tour of the Hubble Telescope. His wish was to spend his retirement years in Hawaii."

After receiving his Ph.D., Mark was hired by Texas Instruments, based in Dallas, in 1983. Many of us know Texas Instruments (TI) for the calculators they make, as the majority of high schoolers within the U.S. are required to buy and use one during advanced math classes. Before its more modern calculators, in the 1980s, the company was working hard to start a nano-electronic program. As Mark described in a 2019 video interview with the National Nanotechnology Initiative: "My charge was to explore the limits of electronics." With Mark's arrival, the TI researchers hoped that the program could break new ground in nanotechnology. They had no idea how right they were about to be.

Creating Quantum Dots

At TI, Mark and his team began looking into developing specialized technologies to make very small devices where phenomena from quantum physics become important. These devices would use tiny crystals of semiconductors, like silicon, in order to create electronic devices. In a later paper, published in 1987, Mark coined the phrase "quantum dot" to describe devices constructed so that their properties were isolated from the macroscopic world.

A CNRS Research Scientist at the Laboratoire de Physique des Interfaces et des Couches Minces (LPICM), at the Ecole Polytechnique in Palaiseau, France, and the writer for the *Nature Nanotechnology* obituary for Mark, Laurie Calvet, explained:

"There are really two large fields of quantum dots. One is based on chemically synthesized nanoparticles, the other is based on confined semiconductors, the type Mark Reed demonstrated. The first is the one that has made such a huge impact on society—most notably because of quantum dots screens, photovoltaics and biosensing." Quantum dots are sometimes called "artificial atoms," because they behave like individual atoms, due to their discrete energy levels.

Mark focused on the electronic properties of quantum dots in confined regions of semiconducting material. In quantum mechanics, there is a rule called the Heisenberg Uncertainty principle. As Mark explained in a 1993 article for *Scientific American*: "The position of an object (an electron, for instance) and its momentum, cannot both be known to arbitrary precision." Quantum mechanical events are then formulated as the probability of an event occurring. This means that researchers can know where an electron is to a high precision, the knowledge of how it is moving will have greater uncertainty. For Mark and his team, the Principle could actually be useful in creating a working quantum dot. "As an electron is more closely confined, its momentum must be more uncertain," Mark wrote in the article. "This wider range of momentum translates to a higher average energy." If an electron is placed in a thin enough layer, the amount of energy necessary to break out of this layer would be so large that it effectively becomes confined. An electron confined to a two-dimensional plane has very little chance to enter a third dimension and an electron confined to a 0-dimensional space has very little chance to move in any dimension. The probability to "leak out" of the confined region is not zero, however, and it is this "leaking out" that allows the properties of quantum dots to be measured using very finely tuned electronics. This initial research helped bring about new directions in condensed matter physics, eventually moving the field towards developing quantum computing technologies using semiconducting devices. Notably, researchers showed how to realize "coupled quantum dots," where electrons have a probability to be in either dot and/or two electrons in quantum states formed by a mixture of the states of each dot, known as entanglement.

Quantum-confined systems in semiconductors can also be used to realize an optical quantum dot, which can also be applied to quantum computing. As discussed in Chapter 3, electrons and atoms can absorb a certain amount of energy, which they then release in the form of light. The difference between the electron's ground state and its excited state, in terms of energy value, is related to what color light is emitted. A higher energy level will cause more of a yellow or green fluorescence, while a lower energy difference emits red light. For optical quantum dots specifically, researchers can use high-powered lasers to activate an optical quantum dot, which, when carefully engineered, can realize one photon at a time. Because of this highly controlled photon release, these quantum dots make for excellent devices to use in different types of quantum computers.

It's important to note that Mark did not discover the quantum dot. The notion of confinement to a zero-dimensional space was first considered in 1980 by the Russian physicist Alexei Ekimov, who was studying glass crystals. Mark's demonstration of quantum dots in semiconductor nanostructures also relied on the development of researchers at Bell Laboratories (see chapter 6) and IBM creating 2-dimensional "quantum wells." These wells acted similar to the quantum dots, with the electrons being confined and emitting light, but in a 2-dimensional plane instead of a zero-dimensional plane. Developing the quantum wells, researchers believed that it was impossible to get the devices any smaller and on a one-dimensional plane. These quantum wells seemed to work just fine, albeit they were extremely complicated to make, as they were built "epitaxially" (a great Scrabble word), meaning one layer was deposited at a time. This made it more difficult and expensive to create identical quantum wells. Seeing a potential opportunity, Mark and the other TI researchers found a way to create identical wells and make them into 0-dimensional dots.

The team used a process called photolithography, or lithography, to essentially "carve" the dots into pillars. The lithography process uses light to produce tiny patterns on thin, nano-films that can be put on a surface, such as a silicon semiconductor. The process also uses etching and deposition to combine the pattern on top of the surface in one layer. As Mark elaborated in the *Scientific American* article: “My coworkers and I...made the first lithographic quantum dots in 1987. We cut slabs of quantum-well material into pillars by means of advanced etching techniques, similar to those used in the fabrication of the state-of-the-art integrated circuits.” In using lithography, Mark essentially created a new way to make electronic quantum dots with more physical isolation in pillar structures.

The original process however was sloppy, as calculating the size to energy-level ratio was not easy. “When we first tried to make quantum dots, no one knew what the correct sizes might be, but many attempts ended in failure,” Mark stated. “But early in the morning of August 20, 1987, as I was preparing to give a talk at a conference on quantum-well devices, my colleagues called to say that they had successfully measured a quantum dot. I rushed to the hotel fax machine just in time to see the data printout, showing a rich harmonic series of energy levels. An hour later, I had rewritten the finale of my talk, the hotel staff had transformed the fax into a viewgraph and I delivered the news.” Thanks to his team's work, Mark was able to create a sort of key for how the size of the dot related to its energy levels. This key would be monumental in helping other scientists use electronic quantum dots for other nanotechnology devices.

While many researchers use electronic or optical quantum dots, they can also make chemical quantum dots using different elements. Though size is the main variable for the dots’ energy levels, structure (hollow vs. solid), composition, and shape can also affect the output. As many of the original dots (and some today) were made with heavy metals, they were toxic to use in industries like medicine. As early as the 1990s, researchers at Fujitsu Laboratories in Japan were looking to make these dots out of organic material, making them safer to use. Today many dots are made out of carbon, graphene, or even curcumin, leading to more applications.

In the life sciences, chemical quantum dots offer many opportunities. They can be used as biosensors, tracking different proteins or bacteria in the body. In a 2022 paper in the *Journal of Chemistry Technology & Biotechnology*, a group of researchers looked at the possibility of using quantum dots to track COVID-19 infections, as well as stop viral infections. Using a "smartphone-based quantum barcode," the researchers were able to track COVID infections in real-time using the quantum dots. As the dots can be paired, or "tagged" onto a virus or viral cell, the researchers could track how the virus spread. The researchers posited that the dots could also be used to stop viral infections, as "functional QDS can disturb the virus-host cell recognition by interacting with the spike protein of CoVs." Other dots could even deliver anti-virals throughout the body. In a separate experiment, the researchers found a 44% and 59% inhibition rate for virus replication using two different quantum dots made of carbon. Of course, there are still kinks to work out, such as the challenge of the dots' low solubility in the body, which makes it hard to release the drugs effectively. There are also many ethical questions about the right to have trackable nanotechnology in someone's body (many science fiction movies show us why this could be bad) and what sort of regulations would be needed. Though research is still ongoing, quantum dots may one day shift how viruses and even pandemics are dealt with, saving millions of lives.

Chemical quantum dots can also be useful in many different electronic devices, like solar cells. "They can be manufactured in an energy-saving room-temperature process; they can be made from abundant, inexpensive materials that do not require extensive purification as silicon does; and they can be applied to a variety of inexpensive and even flexible substrate materials, such as lightweight plastic," explained one article. The most common use for quantum dots is in electronic displays, such as TV or computer screens. Compared to LEDs, quantum dots offer a longer lifetime, lower power consumption, a lower manufacturing cost, and purer colors. Because of these benefits, many companies are looking into switching to quantum dot technology. In 2015, Samsung and LG made the leap when they launched their QLED TVs and displays, still being sold today.

Since Mark and his team were working specifically on semiconducting quantum dots, which is just one of the possible technologies that can be used for developing a quantum computer. As Calvet stated: "Microfabricated quantum dots opened up an entirely new type of physics to explore at low temperatures. This started in 1987 and continues today where they have been explored very intensively for quantum computing. The most promising results are from Intel which claims to have made quantum dots using the same technology as silicon chips. This is the holy grail—superconductors have so far shown the most impressive demonstrations but they tend to be much larger than what you can make on silicon." In quantum computers, optical quantum dots can be used for multiple processes, from helping photonic qubits to assisting in laser stability. With these many applications, it should be no surprise to see optical quantum dots being relied on more heavily in the next generation of our computers and other devices.

For those worried about the environmental impact of using specifically chemical quantum dots, many researchers are working to make the production eco-friendlier. To that end, researchers from the Institut National la Recherche Scientifique (INRS) have found a possible method to create carbon quantum dots from beer waste. According to the lead researcher and INRS professor Dr. Federico Rosei: "The use of spent grain highlights both an eco-responsible approach to waste management and an alternative raw material for the synthesis of carbon quantum dots, from a circular economy perspective." As the brewery waste is also high in nitrogen and phosphorus, it removes the need for purified chemicals, allowing a much cheaper manufacturing process. To create the special carbon quantum dot, the researchers microwaved the grain products and then used the carbonized power to develop a prototype quantum dot. The researchers planned to use the prototype to measure the concentration of heavy metals and other pollutants in the water. As Rosei explained to Kenna: "typically you would functionalize the quantum dots to make them sensitive to certain contaminants, which would attach selectively to the functional groups." While science and (drinking) alcohol do not always mix, when they do, it can create unique and interesting results.

Mark's Yale Years

At TI, Mark wasn't yet thinking about all the potential applications for quantum dots; instead, he was busy developing other nanodevices as part of the nanotechnology boom of the

1980s and 1990s. “At TI, we all were young and excited about what we could do,” explained former co-worker Keith Bradshaw on the virtual memorial page. “He [Mark] came to my desk one day (I had the prototype TI e-beam machine and was pushing its envelope). He had an idea for a quantum resistor...We laid out the mask layout on some grid paper I had in crayon. We changed the design from circular to square to make it easier to align, and it worked. Earlier we had made a containment grid for his first quantum dot paper. He was very generous and had a wonderful personality. I still have the grid paper design. We wrote the code on the paper for the machine to use (the crayon lines have faded a bit).” Mark also spent time with his TI coworkers at the dive club, where he eventually trained to become a divemaster. He would continue to do various dives like shipwrecks or wall dives in Belize or Hawaii for the rest of his life.

1990 saw a significant change for Mark as he joined Yale University’s faculty. “For more than 30 years at Yale, Reed was extremely active and continued his streak of ‘firsts,’ including the first conductance measurement of a single molecule, the first molecule transistor, and the development of CMOS nanowire biosensors,” stated the Yale University obituary.

During his time at Yale, Mark met his wife, Elizabeth, on a first-class red-eye flight from Hawaii to Chicago. According to Elizabeth’s account for *Nature Nanotechnology:* “We met on an airplane returning home from Hawaii. He told me he was ‘just an engineer’ which was his go-to response when asked what he did for a living.” With Mark’s business card in hand, they parted ways in Chicago. They married in 1996 and continued to enjoy traveling together around the world. However, their favorite place to spend time away from home was the Hawaiian Islands. Elizabeth recalled seeing Mark’s laboratory for the first time:

> “On my first visit to his lab at Yale, he casually took out a test tube, put a speck on a slide, and put it under a microscope. He simply said, ‘this is what I do, it is a circuit, it can repair itself.’ Then he simply stepped back and smiled. He was able to make everyone feel comfortable around him in both professional and in social settings. He would never brag about his

accomplishments but rather praise and credit those who worked with him, his students, and his collaborators. I believe that his greatest legacy will be his students who will continue to carry on the research that began in 'The Reed Lab' at Yale," Elizabeth said in the *Nature Nanotechnology* obituary.

Besides being an excellent communicator and mentor, Mark's noted ambition and never-fearing personality helped him achieve many accomplishments. He held 33 U.S. and foreign patents on quantum effects and molecular devices. He received *Fortune Magazine's* "Most Prominent Young Scientist" Award, the Kilby Innovator Award, the Fujitsu ISCS Quantum Device Award, and the IEEE Pioneer Award in Nanotechnology. He was an American Physical Society Fellow, an IEEE fellow, and a Fellow of the Canadian Institute for Advanced Research. He also served as the editor-in-chief of both *Nanotechnology* and *Nanofutures* journals. He also co-founded the *Nanofutures* journal itself. In explaining the difference between science and engineering, Mark stated in the 2019 video interview: "Science explains what is, engineering creates what never was."

Elizabeth recalled to Kenna that "Mark had an amazing work ethic. He was very committed to his research, students, and teaching. I would often find him corresponding with students and preparing lectures into the early hours of the morning and late into the evening, on weekends, and on vacations. In the 26-plus years we were married no matter where in the world we were (home, travel, or vacation) Mark was up at the crack of dawn corresponding with his students long distance. Although it was difficult for Mark to tear himself away from work, he did manage to take breaks to watch a scheduled passing space station or cosmic event, sports on TV (sailing, football, golf) or watch a good science fiction or mystery movie. Of course, when we traveled together, we both enjoyed golf, scuba diving, and sightseeing. However, at home in Connecticut, I had to coax him to take breaks from his work to get outside to do participate in a sport. To Mark's surprise, I submitted a membership application to the local men's senior golf league. He was glad I did this and enjoyed being out on the golf course playing on and off the league. Fellow golfers recall him as fun-loving with a larger-than-life jovial laugh."

Back at Yale, Mark was not one to let his scientific accomplishments get to his head. He focused on making sure to support his colleagues and students. In an interview with Kenna, his colleague, professor Hong Tang said: "He was half a generation above me. In that sense, he looked over my career growth, for the past 16 years."

Fellow Yale professor Fengnian Xia echoed similar remarks to Kenna about Mark's role as a mentor to his colleagues: "Mark was easygoing, humble, generous, and sharp. He is a great mentor and many Reed group alumni are currently playing important roles in both academia and industry. In my case, when I just joined Yale as a junior faculty member, I met Mark in the Becton hallway of the fifth floor late at night multiple times. We talked about many different topics including research, life, and the Yale EE department. His encouragement and guidance are critical for my Yale career."

Many of Mark's colleagues and students described Mark as a cowboy, due to the signature white cowboy hat he always wore. Tang even showed Kenna the hat, bone color with metal beads and feathers, remarking that Elizabeth had given it to Tang as a reminder of his friend and colleague. Elizabeth shared with Kenna that "Mark was a man who believed in living life to the fullest on his own terms. He loved being nonconventional in his work, personal life, and in his style of dress. He had an affinity towards the Southwestern style of dress. He felt most comfortable in jeans and cowboy boots. Not your typical image of an Ivy League professor. I believe that his casual southwestern style of dress was another reason he was so approachable, and his students felt comfortable around him. I laugh when I hear his colleagues refer to him as a 'cowboy,' He did not embrace anything resembling a cowboy's life except for the clothing and relaxed attitude. He just loved wearing southwestern clothing and jewelry."

With the impacts of the COVID-19 pandemic, Mark began looking at nanotechnology's benefits for the health sciences. "When COVID started, he had the plan to find a way to detect human immunity against COVID-19 using his nanowire technology," explained one of his postdoctoral researchers, Mohsen Nami in an interview with Kenna. "But no one was allowed to go to work starting in March of 2020. He told me we need to do something. People were dying all over the world. Our group was the only group that could potentially do something to determine human immunity against COVID in less than 20 minutes. Even a small step may have a big impact on society. While every day 200-250 died in Connecticut and there was no vaccine, Mark came to the lab and worked on the sensing setup his students developed for the past 10 years. I saw that it was not fair for a 64-year-old man to come into the lab every day while I am in quarantine at home. I joined him. He said, 'you do not have to come, you have a little kid and anything can happen,' I told him. 'This is my honor to help you. I owe this country and these people.' Together we began working on his ideas and setup."

Mark's research impacted the scientific community on multiple levels. "Mark was able to work in several fields AND have top papers in those fields," Calvet explained to Kenna. "He did have a specialty—it was coming up with clever ways to measure and to fabricate devices—and he applied this in a variety of fields. This is truly rare, especially today. We tend to specialize early and then stay in that specialization. You can switch fields but it is hard and very hard to publish when no one in the field knows you. Mark did not have this problem. He made very important contributions to microfabricated quantum dots, to molecular electronics, and to biomedical devices." This versatility in Mark's skill set made him a unique figure within the scientific community, connecting with audiences in multiple fields.

Mark's personality also influenced many of the scientists around him when it came to the culture of research. As Calvet stated: "Every scientific department should have someone like Reed around to cut through the typical bullshit that goes on in academia. I think many of the students who ended up working for Mark had someone been rejected by another lab—typically for some silly reason (probably just some kind of prejudice). For the students it always seemed like getting kicked out of a lab for such and such reason was so dramatic, but Mark always saw through this. I think everyone was in general just happy to be working with someone who was so easygoing. I think scientists should try to be more open-minded in general—both about mentality and about people. Scientists tend to be skeptics more than idealists and optimists. Mark was able to do the seemingly impossible because he was so enthusiastic and optimistic. His outlook is just very different from your typical scientist. He was a breath of fresh air in a stuffy room. This should remind all of us that the way to do good science is by having a diverse group of people."

Mark's Mentorship

Besides being a professor, Mark was also the Director of Undergraduate Studies for the Department of Electrical Engineering. This meant that he helped guide undergraduates with their class choices as well as answer any career questions they might have. "He talked to every incoming student, every graduating student, and every student throughout the year," Tang stated. Mark set a precedent for what it truly meant to be a mentor, making sure he was always available to students and telling them that "rules were meant to be broken." Collaborator Peter Burke of U.C. Irvine stated in the *Nature Nanotechnology* Obituary, "During a research meeting, some undergraduates came in with a question. Mark stopped the meeting for about 10 minutes to help them, clearly bending over backward and excited to answer their questions. Being a mentor to everyone was just second nature to him."

For Calvet, a previous student of Mark's, his mentorship made a significant impact. "The first way that Mark impacted me is that, when I was a graduate student, he never once questioned why I would want to do a Ph.D. in applied physics," Calvet told Kenna in an interview. "Just about everyone else I knew couldn't understand it and would ask me what made a female American choose such an unconventional and seemingly difficult path. During my graduate school years, I often felt out of place—when I went home to see my family and friends, at Yale, where I knew very few other female Americans who were doing the same thing as me. To be clear there were other women in Mark's lab, but they were international students. That helped, and today I feel very close to them, but at the time I think we were all dealing with our own very different types of problems and it was hard to relate to each other. I never felt out of place though when it came to a group meeting or a conversation with Mark. He just accepted you as you were. I heard this echoed also by undergraduate students on the Zoom that we had after his passing. He immediately made people feel at ease—the only thing he asked really is that you 'loved' doing science and if this was the case he was happy to have you around and working with him. In short, Mark gave me the confidence to believe in myself as a scientist like no other person I had met before. No matter where I was, I knew that there was this Professor at Yale who had not one doubt that I could be a good scientist." Calvet used Mark's approach to mentoring in her own career. "When I mentor students today, I am constantly reminding myself of how he would be: 'Slow down, be kind,' I tell myself and then ask them the right questions until they understand better." Both Brian and Kenna experienced the importance of sincere, structured mentorship in their own lives, playing the roles of mentees and mentors at various points themselves.

Mark passed away in the midst of teaching the course "Advanced Electron Devices." His students and research groups were devastated. Former student Adam Wolnikowski credited Mark with convincing him to major in electrical engineering. "Personally, he inspired me to get my academics back on track after a bit of a slump my first two years of Yale," Wolnikowski stated in an article in the *Yale Daily News*. "He shared with me invaluable wisdom on planning a career in electrical engineering. He will be clearly missed and this is truly a great loss for the engineering community at Yale." Others, like Nami, are still grieving Mark's loss. "The day that I heard he passed away was one of the worst days of my life," Nami said. "I couldn't believe it. Even days after that, I thought one day he will still come into the lab and say 'hello,' with his cowboy hat. I've dreamed about him several times. I'm sure I will meet him again and tell him that his ideas worked out."

The 23 Ph.D. students and 14 postdocs Mark oversaw held a special place in his heart. He continued to support them after they graduated by giving career or scientific advice, and even investing in many of their companies. "In tribute to each of his Ph.D. students, he made a photolithographically etched silicon wafer in honor of each thesis," explained the *Nature Nanotechnology* obituary. "The board with these silicon wafers was placed in a prominent position in Mark's office and was an oft subject of conversation during a visit. As a memorial to him, the board will remain at Yale near his former laboratory." Elaborating on the silicon wafer board, Calvet told Kenna: "We were 'his kids' and we are his legacy. I hope that my students will have this same impression of me."

While Mark has passed away, his legacy continues to live on in those whom he worked with and mentored. To honor his work the Institute of Physics (IOP) publishing company has remained their Nanotechnology Young Researcher Award in honor of Mark. Both Kenna and Brian believe the works of former Yale student Alejandro Carrillo summarize Mark's impact very nicely, as Carrillo wrote on the virtual tribute wall: "I still think of him when I look at glass or discuss electromagnetic radiation and materials with colleagues in my professional life. The last thing he ever said to me was in the Becton courtyard during graduation activities and a group photo of the EE class of 2016. I was thinking anxiously about my undefined future post-Yale and probably looked pale and nauseous. He looked at me with concern, came to my side, and asked me, 'are you okay?' Mark Reed was brilliant, kind, and a light to thousands of students. He is deeply missed by myself, and every other student I have spoken to in my cohort."

Mark's loss was felt by more than just his immediate family. Elizabeth told Kenna in an interview, "Mark's passing went from me privately grieving with family members to being public within a matter of just a few days. The Yale University community rallied to create a series of Zoom memorial services. We were still in COVID lockdown in America, and around the world, so it was not possible to arrange an in-person memorial service. Therefore, the series of Zoom memorials arranged by Laurie Calvet and other members of the 'The Reed Lab' was a wonderful unconventional solution to allow people from around the world to gather and grieve together. Mark would have appreciated what his lab members orchestrated. It was comforting and inspiring that so many people gathered to pay their respects to Mark and his legacy. I am grateful for Laurie's (and others') quick response to organize the memorials."

Yale University was also helpful in supporting Elizabeth through this time. "I was given access to Mark's office to gather his personal effects at the University," said Elizabeth. "After participating in the Zoom Memorials, I was very aware of how close his students and fellow professors felt towards him. Many of them referred to him as a 'father' figure in their lives; never thinking that there would come a day when they would be unable to reach out to him for assistance or guidance. I feel that Mark was the common thread amongst all of us. As such, I gave, and also left some personal items in his office for students and faculty to have as mementos." One of these was Tang's bone colored hat, which he proudly displays in his office to this day.

"Mark passed peacefully in his sleep in our home in Connecticut from heart failure, not Covid related," Elizabeth clarified. "After his passing, I was asked by one professor if he would be interred in the Yale cemetery as some professors had chosen to do. However, Mark's wishes were that he be cremated, and his ashes are released into the Pacific Ocean off the shores of Hawaii. I did this on the anniversary of his death in a Hawaiian Sea Ash Scattering Ceremony off the coast of Kauai. It is believed that the sea off the Kauai coast is where all souls come to this earth and when we die, they return to the sea. To quote Heinrich Zimmer, '*Limited and immortal, the waters are the beginning and the end of all things on earth.*'"

Elizabeth believes that: "Although Mark is physically no longer with us his spirit lives on in each person who was touched by his presence during his lifetime. His students will be part of his legacy. Using the nanotechnology research foundations learned in the 'Yale Reed Lab' may they always remember Mark's mantra 'rules are made to be broken' to reach new heights in their contributions to the advancement of nanotechnology."

Chapter 10

Deborah "Debbie" Jin - "The Goddess of Fermions"

A loud *clink* is heard as the metal bat projects the softball high into the air. It's hard to see where the ball goes as it seems to dive into the bright Colorado sunset. Cheers erupt from the stands as players hustle from one base to another. In the middle of the commotion, the batter, Deborah Jin, runs as fast as she can to first base, her number 35 bright red jersey streaking across the field. Her fellow players and research colleagues cheer her on, as her face shows complete determination. She is going to get to that base, and no one can stop her.

***

Perhaps Kenna was a bit biased in suggesting Deborah "Debbie" Jin for the final "giant" of this book, given that JILA (a leading physics research institute created by the partnership between the National Institute of Standards and Technology, or NIST, and the University of Colorado Boulder (CU Boulder)) is where they both worked/work. When Kenna first arrived at JILA in 2020 as JILA's science communicator, she was told only a few handfuls of stories about Debbie, who had tragically passed away in 2016. Each story made Debbie seem more mysterious and legendary than the last, and after some research, it wasn't hard to understand why. Debbie not only made significant strides in advancing our knowledge of how atomic interactions work on a quantum level, but she also was a role model and mentor for many students, all now doing research across the world. She pushed the scientific community to become more aware of work-life balance with families, all while using freezing temperatures to unlock new quantum discoveries at the forefront of science. Her genius and her compassionate spirit make her an inspiration for curious minds of all levels. "She was an incredible human being and a role model for everyone," explained JILA and NIST Fellow Jun Ye, a colleague of Debbie's.

Author's Note: Like Betty's story (chapter 7), it is important that Debbie's story be shared in schools and in books like this. And like Mark's story (chapter 9), Debbie's is still unfolding, as many of her projects continue to this day. As Debbie was a rather private individual when she was alive, Kenna and Brian will do their best to honor her request for privacy in this chapter by focusing on her later accomplishments in physics and less on her personal life. We're both extremely grateful to her friends and colleagues who have shared their stories about Debbie with us.

Because physics, especially quantum physics, is known for its low number of female researchers, it can be hard to find appropriate role models in this field. According to a 2020 survey by the American Physical Society (APS), only 27% of the bachelor's degrees in physics were awarded to women in 2020 and only 20% of PhDs were awarded to women in the same year. Additionally, women comprise fewer than 20% of faculty positions in physics at American universities, according to the American Institute of Physics (AIP). Some experts have event posited that it won't be until *the next century* that the number of women in physics will reach 50%. Clearly, these numbers show that there is much need for improvement in this field. Stories like Debbie's can not only motivate the women who are currently in these fields but also inspire others to consider physics as a future career by seeing someone who looks like them being successful there. Indeed, what makes Debbie's story unique is that she was not just a woman in quantum science, she was a pioneer. "The top scientist in the world, in this field, is a woman," explained John Bohn, JILA Fellow and Debbie's husband, in an interview with Kenna. To see someone like Debbie leading this field of quantum physics is rare, and further solidifies her status as an absolute giant.

Born in 1968 in Stanford, California, Debbie was one of three children in a family of scientists. Her father was a physics professor and her mother was an engineer. Debbie grew up in

Indian Harbor Beach, Florida. In a *Science* obituary, Ye and JILA Fellow Cindy Regal, both of whom had worked with Debbie, wrote: "Growing up, Jin had played the violin, was most drawn to mathematics and was helped along in her scientific interests by discussions with her father." Debbie originally wasn't interested in physics, but only became interested after a summer job at the Goddard Space Flight Center in Maryland. After that, Debbie went to Princeton University and got a B.S. in physics in 1990. According to a 2017 Princeton Alumni obituary, Debbie was already interested in low temperatures. "Majoring in physics, for her senior thesis, Jin built special refrigerators to cool detectors for cosmic-ray observatories in Antarctica," the article stated. After receiving her bachelor's degree, Debbie decided to go to the University of Chicago for graduate school. There she would not only meet her future husband John Bohn but also look at fundamental building blocks of the physical world: fermions.

Initial Work on Fermions

All quantum particles in nature are sorted into one of two categories: fermions or bosons. Fermions are the constituents of matter: things like electrons, neutrinos, muons, and of course quarks. Large, composite particles such as atoms or molecules can also be fermions. According to Brian DeMarco, a professor at the University of Illinois and Debbie's first graduate student: "The rule for how to classify particles (fundamental or composite) is by their total spin. Spin is what it sounds like—how much rotation they have. The earth spins once every 24 hours. Particles have a similar measure. Fermions have a half-integer of spin—1/2, 3/2, 5/2, etc. Bosons have an integer number—0, 1, 2, 3, etc. This is a more general organizing principle that covers fundamental and composite particles." The fundamental class of bosons includes force-carrying particles like the Higgs-boson, photons, or gluons. Both types are still being studied by scientists to hopefully answer some big questions, like the origins of our universe.

What makes fermions unique compared to bosons is that identical fermions cannot be in the same quantum state. Debbie's own version of this is that fermions are “individualistic,” whereas bosons are “conformists.” Because of this, fermions can be difficult to study, which is why many physicists, like Debbie, engineer special systems for controlling them, usually at temperatures close to absolute zero Kelvin (or -273.15 °C or -459.67 °F).

It was at the University of Chicago that Debbie fell in love with fermions...and her husband, John. “We were graduate students together in the same year,” John told Kenna, “One of my earliest memories was that we were in an electromagnetism class together. And we were asked to do some sort of laboratory project. She was working with her partner and I was working with my partner. And her approach was to measure capacitance, a simple thing. So, they built these things out of tin foil and measured the capacitance of the system. During the project, Debbie and I were talking back and forth about it. When they wrote the report about it, which they had to present to the whole class, their story was that capacitance was really important, as Benjamin Franklin used to use it to kill turkeys. So, for me, I realized that she thought that out, and it was a great introduction to her. She clearly had a wicked sense of humor.”

For her Ph.D. thesis, Debbie was advised by Dr. Thomas Rosenbaum, an expert on quantum dynamics of specific materials, such as the electronic and magnetic properties, at temperatures near 0 Kelvin. Currently, Rosenbaum is the President of the California Institute of Technology. Under Rosenbaum, Debbie studied heavy fermion superconductors. According to the Princeton article, these were “an exotic variety of materials that conduct electricity without resistance.” These heavy superconductors included elements like uranium and platinum.

In 1995, Debbie received her Ph.D., and she and John began to figure out their next move. In academia, for those who might not be familiar, once someone gets their Ph.D., if they want to stay in academia, often they will also try to get their postdoctoral degree or “postdoc,” which usually runs for two to five years. These positions can either be contract based or be awarded as a fellowship. In the science community, it’s commonly frowned upon to stay at the same institution to get your undergraduate,

Ph.D., and postdoctoral degree (though it can be successfully done and has been). This unspoken rule allows researchers to study in new laboratories as well as expand their networks. "We were looking for postdoc positions," John said. "We were already married at that point. We had gotten married a year before defending our PhDs because we didn't want both things to happen at once. She was the smart one, she was the one who decided on that." In looking around for postdoc positions, Debbie and John realized they may have a difficult time finding somewhere. "Our one criterion was 'we're going to go to the same place no matter what we do. We're not going to separate,'" John added. "That was a big deal back then, it still is a big deal now."

If searching for one job wasn't already hard, searching for two is nearly impossible, especially in niche fields like quantum physics, where there are fewer academic positions in general. The 50th Anniversary book for JILA called *JILA: the First 50 Years* calls this the "two-body problem," an homage to both the "three-body problem," and the "many-body problem" in physics. (Note: there is also of course a "two-body" problem in physics which has nothing to do with married researchers) Most married scientists try to find jobs as close as possible to each other, even within the same institution on occasion. Many institutions, like JILA, NIST, or even Columbia University, often try to make accommodations for married scientists. That still doesn't make it easy, especially for early-career researchers like postdocs. "A young researcher is lucky to get an academic position," explained Bob Nelson, the public affairs officer for Colombia University in an article for *Science*. "Two is definitely unheard of."

While finding two research positions is challenging, the overall number of academic couples may be higher than you'd expect. A 2017 *Nature* article cites a "report by the Clayman Institute for Gender Research at Stanford University in California which collected data on around 9,000 faculty members at 13 universities, 36% of US faculty members were part of an academic couple." The *Nature* article also mentioned that many of these couples (38%) worked in the same department, which could lead to future research collaborations.

This would be the case for Debbie and John, as they would collaborate on many projects together (according to John, Debbie said that he had written more papers with her than any other theorist). Before then, they had to find jobs.

"We had extraordinary luck at the time," John said. "I knew Chris Greene [a former JILA physicist]. He said he would be happy to take me." After John's offer, he and Debbie began scouring the Boulder, Colorado area for postdoc positions she could get. "She found this 'Eric' guy whom she didn't know from just a hole in the ground," John added. "But she applied, even though it was not the field she had previously studied in." That "Eric" guy would turn out to be Dr. Eric Cornell, a NIST and JILA Fellow and the winner of the 2001 Nobel Prize in physics. Just as Debbie was joining Eric's team at JILA, they were busy at work discovering the exact subject worthy of the Nobel: the Bose-Einstein Condensate.

Joining JILA

For a researcher, JILA offers a unique place to work as it is a half university, half government laboratory. Originally called the Joint Institute for Laboratory Astrophysics (JILA), the Institute was officially established in 1962 as a joint institute between the University of Colorado Boulder (CU Boulder) and the National Bureau of Standards (NBS, later to be called NIST...see chapter 7). JILA was originally founded as a place to study laboratory astrophysics, "from studying the Sun's visible light spectrum to developing reflective mirrors," Kenna wrote in an article for JILA's quarterly journal *Light & Matter*. Over the decades JILA took in more Fellows, or principal investigators (PIs), and expanded into other fields of physics like atomic physics, laser physics, and chemical and bio-physics. In 1994, JILA dropped the acronym entirely as the institute had a much broader scope of research.

Today, around 70% of JILA Fellows (there are about 30 Fellows total) study some aspect of quantum physics. The Institute has also been the home of three Nobel Prize winners: Eric Cornell and Carl Wieman in 2001, and John "Jan" Hall in 2005. Because of its joint status, there is a much more independent culture among the researchers, as many feel they can study almost anything they want to. NIST's Boulder laboratories are also just down the road, so it's easy for researchers to collaborate with one another or get any questions answered. For Debbie, JILA would become the perfect choice. As John remembers it, it took a bit of coaxing to convince Debbie to switch fields of study for her postdoc. Most researchers blanch at this idea because they feel they are entering an entirely new arena, with no previous background. Debbie felt the same way. "Eric wrote her this long letter," John said. "She showed it to me one time. The thing I remember about it is that he said 'you don't have experience in lasers, but you do have a CD player, don't you? There's a laser.'" After some further deliberation, Debbie decided to take the job at JILA, even if it meant learning an entirely new subfield of physics.

Just before Debbie arrived at JILA, Fellows Eric Cornell and Carl Wieman were busy at work trying to produce the Bose-Einstein-Condensate (BEC), a new state-of-matter. As discussed in chapter 4, the BEC was a theory published in 1924 that described the quantum dynamics of atoms that get closer together at extremely low temperatures close to 0 Kelvin. Both Cornell and Wieman wanted to show that the theory could be validated in a laboratory. Using special magnetic and laser traps, the team was able to create a BEC made from rubidium atoms.

The process began with the rubidium atoms being a room-temperature gas. From there the atoms were held in a laser trap and constantly bombarded by photons (light particles) on every side via lasers. In a 1998 article for *Scientific American,* Cornell and Wieman explained that "the laser light need not be intense, so we obtain it from inexpensive diode lasers, similar to those found in compact-disc players." The atoms were then cooled to around 10 millionths of a degree above 0 Kelvin. After the atoms were trapped, the researchers turned off the lasers and turned on a magnetic field to keep the atoms in place. Then the atoms were cooled even further. Cornell engineered a way to help keep the density high enough at the ultracold temperature in order for the BEC to form. The first BEC lasted around 15 to 20 seconds and was made of 2,000 rubidium atoms. To preserve the BEC, the team took a "shadow snapshot" using a flash of laser light to record the shadow of the condensate in an image. The researchers could then use the image to measure the velocities of the atoms inside the BEC and make different distribution images. This experiment could then be repeated at different temperatures to see different distributions in the images. “It's kind of the triumph of quantum mechanics,” Cornell told Kenna in an interview, “Bose-Einstein Condensates aren’t all that small, they're a fraction of a millimeter across, which is a relatively large scale compared to other measures in quantum physics. So, the only reason why the quantum mechanical effects are visible over those larger distances is that they are extremely cold.” Thanks to their hard work, the team's discovery could bring quantum phenomenon to a visible level. As Cornell and Wieman wrote in the *Scientific American* article: “quantum-mechanical waves extend across the sample of condensate and can be observed with the naked eye. The submicroscopic thus becomes macroscopic.”

BECs may also play an important role in the future of quantum computing. BECs provide a unique platform for quantum computing, as they allow for the manipulation and control of large numbers of qubits (quantum bits) in a single, coherent state. According to Ye: “The technology that created BEC is what drives quantum physics (including quantum computing based on neutral atoms) to be so vibrant today.” Recent research at Georgia Tech University is looking to see if BECs can be used as a communication channel between two quantum computers. Professor Chandra Raman explained in a Georgia Tech article that: “You can imagine having a drop of dye that is concentrated at one point in space. Through diffusion, the dye molecules move throughout the water, slowly spreading throughout the entire system.” Because qubits within a quantum computer can break down or “decohere” due to readout and environmental noise, experts are worried that with more qubits in a quantum computer, the more likely decoherence will happen. Raman and his team hope to avoid this by possibly using a BEC. However, the research is still ongoing, and it’s too soon to see if a BEC can be used in this way.

Back in 1995, Debbie had arrived in Boulder Colorado, just as the BEC was being discovered at JILA, with no background in that type of quantum physics. "She had done her graduate thesis in solid-state physics," Cornell stated. "I think it's kind of interesting that her graduate work was in exotic fermions and then later she sort of became the goddess of fermions." Because Cornell and Wieman were looking at bosons and not fermions, Debbie would be learning about an entirely different class of atomic species while on the job. "She showed up just a month after we saw the BEC, not knowing anything about our particular kind of experiment," Cornell added. "But being a very sharp person who was good around the lab, she was very quick on the uptake. As she was a postdoc it was her job to run the laboratory, so it didn't take her very long to assume that position. By the end of her two-year postdoc, she was really quite an expert. So, JILA hired her." One of Debbie's former colleagues, Zheng-Tian Lu, who later invented a powerful new technique of archeological dating by counting cold atoms one by one, posted later on Debbie's online memorial page that he tells his students about Debbie learning difficult quantum physics from scratch and "went on to make one after another fantastic breakthrough," in order to inspire his own students.

Debbie was not only making an impression within the NIST and JILA communities but also in the international physics community. In 1996, DAMOP (the Division of Atomic, Molecular, and Optical Physics), part of the American Physical Society (APS), invited Cornell to go to its large annual conference to discuss the BEC. Cornell declined the invitation and sent Debbie instead. She would be presenting the new BEC findings to at least 1,000 curious physicists. No doubt this would be terrifying for many, especially if you just learned the fundamental concepts only a year before. "She kills it," John said, smiling. "She is the best of the four speakers. She just rose to the challenge." This conference would be only the beginning of Debbie's reputation as a pioneer in quantum physics, and an extraordinary speaker.

In 1997, Debbie became a NIST physicist and JILA Fellow, as well as a professor adjoint at CU Boulder. She opened her own laboratory and decided to use similar methods to Cornell's and Wieman's work on the BEC, just with fermions. "This was a good idea and a natural extension, I thought, because fermions are so very different from bosons," said Cornell. "Both of them start to act up when things get cold, but fermions are hard to control. Debbie had the advantage of being here at JILA, where we already knew how to do a lot of these things. While other people were still trying to see Bose-Einstein's condensation, she had had the advantage of having worked on it for two years. So, she knew a lot of the tricks, and had some good ideas for how to extend it to fermions which she made happen in an amazingly short amount of time."

Debbie's first graduate student was Brian DeMarco, who would help her achieve some significant discoveries and be the first of many students she would mentor. DeMarco had briefly been a graduate student under Debbie's leadership when she was a postdoc in Cornell's lab, so he had experience working with her already. DeMarco joined Debbie's lab in a rather roundabout way that isn't an uncommon occurrence within research: his advisor, Cornell, suggested he change groups. "One day, Eric called me up to his office," DeMarco recalled to Kenna in an interview. "He suggested that my current research project was likely a dead end and that I should consider switching advisors and help Debbie to start her group. I agreed that this could be a terrific opportunity, so I met with Debbie to hear about her plan. She had a clever idea for how to enable fermions to collide at low temperatures, which was one of the big challenges. Debbie's strategy was to trap two spin states of potassium 40. Debbie tasked me: 'tell me if you think this will work.' I thought about it and told Debbie that this strategy could succeed, but I asked 'what are the big gaps, Debbie? If I'm going to switch groups, then I need to know the barriers ahead of us.'" After Debbie and DeMarco discussed the project further, DeMarco officially switched research groups and began to work with Debbie.

It was just the two of them in the laboratory for several years. There were some initial setbacks with Debbie's projects, as the laboratory space still had to be remodeled to her specifications. DeMarco remembered working on her office floor during these renovations, building the lasers and electronics that they would need for their experiments. "We worked together every day, planning, designing, and building equipment," DeMarco said. "We were very compatible as researchers and became good friends. I met my future wife, Ariana, while I was working with Debbie. She was Debbie's age, and Debbie, John, Ariana, and I became close and spent a lot of time outside of work together." The similarities between DeMarco and Debbie allowed them to work synergistically in tackling the difficult problems around fermions.

Debbie was just one of many scientists interested in fermions during this time. "In '97 she moves to fermions," John explained. "And she's not the only one. There are other people who test the water and say 'Oh this is going to be a big deal.' But there's that technical challenge where fermions avoid each other so they don't run into each other. So, all these well-established and well-funded giant research groups with a dozen people get heavily involved." Even scientist Wolfgang Ketterle, who would share the 2001 Nobel Prize with Cornell and Wieman, was interested. "But they all got scooped by Debbie and Brian working alone in the lab," John added. "Scooping" was the absolute perfect word for what Debbie and Brian did, as using Cornell's and Wieman's BEC techniques, they were able to achieve the first Fermi degenerate gas before anyone else could.

Fiddling with Fermions

While fermions may seem like an alien concept to most of us, life wouldn't happen without them. "Most of the important stuff in the world is fermions, like electrons," Cornell stated. "Electrons carry electricity and this can cause things like chemistry to happen. So, by making an ultra-cold fermionic gas, Debbie was making a sort of model system for the electrons." Using the techniques that Wieman and Cornell taught her, as well as techniques that she invented herself, Debbie was able to create a dense, ultra-cold cloud of gaseous fermions. With this new dense phase of gas, she could now see the phenomenon of quantum degeneracy.

In chapters 3 and 9, we discussed the discrete energy levels, or wavelengths, of electrons and other particles as a part of quantum mechanics. Quantum degeneracy takes this process one step further. "When the atoms are cooled to a low enough temperature, the indistinguishability of the atoms becomes important," DeMarco explained to Kenna. "And the interference of the overlapping quantum waves determines the properties of the gas. That interference happens in one way for bosons (such as in the BEC), but differently for fermions. For fermions, the waves cancel out so that two particles are never in the same place at the same time. Sometimes I think that it is unbelievable that this cancellation is exactly perfect, but that is our understanding of nature!" The overlapping, degenerate waves created something called a Fermi Sea, named after physicist Enrico Fermi. According to a History of Scientific Women article, the "theorized 'quantum degeneracy' occurred at around 300 nK," for Debbie's fermionic gas (which to say is cold is an understatement!). In using similar techniques to Cornell's and Wieman's, Debbie and DeMarco could show the fermionic interactions happening in real-time. "Now she had quantum mechanical atoms, which are much, much larger than electrons, much easier to see or track or follow. It's kind of like looking into a microscope," Cornell added. "It was astonishing that it happened so soon for her. She started as a professor in 1997 and it really took her only a couple of years to see a degenerate gas."

This was a big deal for both Debbie and DeMarco, and they knew they had to publish their findings right away. "We wrote up our results and sent a paper to *Science,*" DeMarco said. "Those days were a different era, and we mailed a printed copy of the paper with the digital figures on a compact disc. We held our breath while the results were under review. Then we received the first referee reports, wrote a response, and the paper was finally accepted." Thanks to the 1999 *Science* paper, the news spread quickly, solidifying DeMarco's and Debbie's reputations as successful researchers. "I came into the lab one morning after publication," DeMarco added. "And reporters started calling from *The Washington Post* and *New Scientist.* The attention was exciting." As DeMarco explained to Kenna, Debbie was generous with the incoming publicity and shared many of the interviews and scheduled talks with him, allowing him to develop as a researcher and communicator. "It was a great opportunity to grow as a scientist," DeMarco stated. This publicity also helped them to pave their own road in the unexplored terrain of fermionic science. John said: "That really put her on the map. People were struggling and just not getting there and here's Debbie just doing it."

2001 was a big year for JILA, CU Boulder, and NIST as Cornell and Wieman received the Nobel Prize in Physics, along with Wolfgang Ketterle. "Wieman heard the news about 4:00 a.m. on October 9 from his younger brother who had checked the internet," said *JILA: the First 50 Years*. "Cornell heard the news from his former thesis advisor, David Pritchard of MIT. An impromptu gathering was held in the JILA auditorium that afternoon. Wieman and Cornell received thunderous applause from a capacity audience and opened a bottle of champagne to loud cheers." Though Debbie arrived after the initial discovery of the BEC, she helped with some of its later test experiments. She did travel to Stockholm with Cornell and Wieman for the Nobel Prize ceremony, watching the medals be given to her former advisors. According to both John and Ye, Debbie never really spent any time worrying about receiving this prestigious award, instead, she remained focused on her research.

Motivated by her previous success, and the success of her mentors, Debbie pushed on in her exploration of fermion dynamics. Her laboratory welcomed a new member in graduate student Cindy Regal given in 2001 DeMarco had graduated with his Ph.D. and moved down the road to NIST. Like DeMarco, Regal found Debbie a helpful mentor and advisor. "She was always one to take any turn in the lab needed to reach her science goal," Regal told Kenna. "She worked hard to make sure there was a deep connection between the data and the results. Her mindset was always to look at a problem from multiple angles before drawing conclusions." With Regal's help, as well as that of postdoc Markus Greiner, Debbie was able to push fermions into a new regime where pairs of fermions would behave similarly to bosons. This regime was created using Feshbach physics, where specific magnetic frequencies were tuned to coax the fermions to pair up with each other. These Feshbach resonances happen when two atoms slowly collide and temporarily form an unstable compound. The compound only stays together for a little bit, so there is a time window for studying these interactions. This research would eventually culminate in the creation of the first fermionic condensate in 2003.

Like the bosons condensing within a BEC, pairs of fermions (also known as Cooper pairs) also condense in a similar manner. With her team's assistance, Debbie was able to implement previous techniques she had learned, and new ones she engineered, to create the first-ever fermionic condensate. Not only was this condensate extremely technical to achieve, but it also gave more insight into different quantum dynamics. "They were able to show the BCS-BEC crossover, which has been theorized since the 1970s, for the first time," explained DeMarco. The BCS theory (named after scientists John Bardeen, Leon Cooper, and John Robert Schrieffer) described superconductivity (the ability to conduct electricity without resistance) as being caused by the condensation of Cooper pairs. The fermion gases Debbie was exploring showed BCS pairing, where the Cooper pairs were weakly coupled to each other. Creating a fermionic condensate, Debbie and her team could manipulate the Cooper pairs to bind tighter to each other, more closely representing bosons. This transition process is called the BCS-BEC crossover, and Debbie and her team were the first to see this happen. "This was a touchstone point for understanding how fermions paired together," Regal explained.

With the advances of creating the first fermionic condensate and seeing BCS-BEC crossover, Debbie's work suggested many big implications. "Consequently, Jin's work opened a new field of research: fermionic quantum simulations," stated a 2017 paper published in *the Proceedings of the National Academy of Sciences* (PNAS). "Jin showed how a precisely controllable gas of fermionic atoms can be used to explore the physics of strongly correlated electron gases as they appear in superconductors and other solid-state materials. This toolset she devised is now used all around the world."

From this work and her other advancements, Debbie won the MacArthur Fellowship (also known as the "genius grant") in 2003. It would be just one of many awards she would receive over her life, as she was also awarded the Maria Goeppert-Mayer Award (2002), the Arther S. Flemming Award (2003), the Service to America Medal (2004), Scientific American's

"Research Leader of the Year" (2004), APS's I.I. Rabi Prize (2005), the Benjamin Franklin Medal in Physics (2009), a NIST Gold Medal (2011), L'Oréal-UNESCO for Women in Science Award (2013), the Institute of Physics Isaac Newton Medal (2014), and the Comstock Prize in Physics (2014), among others.

In 2005, Debbie was elected to the National Academy of Sciences, and for several years remained the youngest member ever to be elected (and is still considered one of the youngest female members to be elected). Debbie never let her awards get to her head, as she was constantly laser-focused on her research. Debbie often told Ye that "things don't change at all for us," when discussing awards and prizes, showing her dedication to her work. Even when she gave birth to her daughter Jaclyn, "Jackie," in the early 2000s, she still found a way to balance motherhood with research, showing the world it was possible to do both.

Showcasing Motherhood

With her "no-nonsense" ambitious spirit, Debbie was not one to let motherhood constrain her career, but she also wasn't going to sacrifice her family for her research. "She was a fantastic mother," John told Kenna. "From the time Jackie was born, we would go to meetings or conferences and Jackie would go along with us. There, everybody could see Debbie interacting with her daughter. That's being a role model."

Because around 80% of scientists over the age of 40 are parents, the combination of scientist and mother (or scientist-mothers as Kenna calls them in her 2021 article for *the Debrief*) is quite common. Yet, seeing these children at conferences or other meetings is rather rare, and gives this general feeling that scientists should keep research and family separate. This can affect a scientist's overall career. A 2021 *Nature* paper cited a survey where "38% of scientist-mothers worldwide said that they had received fewer offers for professional opportunities since becoming a parent, compared with just 13% of scientist-fathers. And nearly two-thirds (61%) of scientist-mothers, compared with 38% of scientist-fathers, said that parenthood had negatively affected their career." So, while there are no official concrete rules against bringing children into a laboratory or office in many institutions, the general consequences seem to advise against it.

But this can be a huge detriment to the general scientific community, as "hiding" the family side of research can often make scientists appear colder, more robotic, and less human. This is especially prevalent in science fiction (such as the *X-Files*) where scientists (like protagonist Dana Scully) must choose between a child and a career, ultimately picking the career and falling comfortably back into these stereotypes (for more on this see Kenna's *Debrief* article about the need for more scientist-mothers in popular culture). "I think there's often the perception that scientists are intellectually brilliant but barely human," DeMarco explained. "Nothing could be farther from the truth. I have a family, and I go home every night and cook dinner for them. Scientists don't talk about our human side often enough."

To make matters worse, many of the institutions that scientist-mothers work at were created by, and built for, men. From a lack of lactation rooms at scientific conferences for breastfeeding mothers to a lack of adequate bathroom stalls (which happened in 2022 at the University of Florida), scientist-mothers have an uphill battle. While many of these institutions have begun to make accommodations as these issues get resolved, more improvements could be made. Scientist-mothers like Debbie also made their own statements against these barriers (often called "the maternal wall") by bringing their children to work with them, directly confronting the stereotypes and societal "rules of the community. As Ye wrote in a 2017 article for NIST: For quite a few years after her daughter Jackie was born, Debbie's criterion for accepting international invitations was whether the destination city or campus was child-friendly for travelers." As Debbie was such a renowned figure in the scientific community, her deliberate choices of conference venues may have inspired many to rethink their facilities if they wanted to have her future attendance. Debbie's purposeful choices around family inspired many, including DeMarco. "Whcn my child was young, I made sure to tell my students about dropping off and picking him up from daycare," DeMarco said. "It is crucial to show that scientists can have a fulfilling personal life."

Debbie also made sure that JILA, her place of work, was also accepting of families. As Kim Monteleone, a previous JILA staff member wrote on Debbie's virtual memorial wall in 2016: "When Debbie and I began planning the Fellows dinner, she asked to make it family-friendly. She wanted the children of the Fellows to feel welcome. Her face lit up when she talked about this, knowing that her daughter Jackie would be involved in the dinner. Debbie brought games and coloring books to the Gold Hill Inn for the kids. How thoughtful! I have very fond memories of working with Debbie. I loved her smile and getting to know her." Monteleone's sentiments were echoed by many other researchers on the memorial wall, as they offered up their own stories illustrating Debbie's importance in having a rich life full of family and research.

The Science of Softball

While it's easy to think that researchers spend all day in the laboratory, many institutions work hard to make sure this doesn't happen. At JILA, the staff host year-round events such as educational lectures, cookouts, movie nights, and more to encourage work-life balance and bolster the sense of community. Historically, JILA has also created various sports groups, taking advantage of the active Colorado culture. In 2004, JILA Fellow Heather Lewandowski (a.k.a. "Coach Lew") organized a JILA softball team. known as the "JILA Monsters" (a play on the Gila monster). The team joined the City of Boulder's Parks & Recreation League program and included John, Debbie, and Cornell as players. As JILA alumna Dr. Carrie Weidner told Kenna, "I played softball growing up, so I was immediately recruited (and would later joke that this was a form of job security for me because if Heather fired me, she'd lose her third baseman). For as long as I can remember, Debbie was on the team as well. A core of folks in JILA (and JILA-adjacent) was on the team with a rotating cast of grad students and postdocs." Jackie played the important role of scorekeeper at these weekly summer games.

With a whole active team, the JILA Monsters quickly became adopted by the entire institution. “John used to write up these elaborate synopses of the games that would go into the *JILA News* that came out every Friday,” Weidner added, though John told Kenna that Cornell was actually the one that began this tradition. Weidner eventually began trading off with John to write these summaries. In *JILA: the First 50 Years,* a particular 2010 softball summary by John reads: “The JILA Monsters split an adorable double header this week against the Tiny Killer Kittens and Thunder Bunnies... In the early innings, runs were hard to come by, one exception being a solo shot over the right-center-field fence by John ‘Bread’ Bohn. ‘Well, I guess I got a rise out of that one’ said Bohn. ‘It was the yeast I could do. It was a crummy result, though; we really kneaded the win.’” This summary gives just a hint at the community spirit of JILA in the early 2000s, something that few other institutions have, even now. Cornell recalled that softball was just one way that Debbie made sure to support her fellow researchers. “She continued to be a role model for the professors younger than her by setting an example of being a good colleague,” Cornell said. “She was always thinking about the institutional health place as well as one's career. I think she was always very cognizant that you wanted to have not just a healthy research program, but a healthy organization. To do this she took her turn as the Fellows Chair and oversaw things and really made the effort.” Debbie made sure to support this community and often encouraged her students to join the softball team or other activities outside of the laboratory, setting the example for a healthy work-life balance.

As the phrase, “science never sleeps” does ring somewhat true, it can be easy for researchers to develop an unbalanced relationship with their work, as many experiments run at all hours. Even if you leave an institution like JILA at 5 PM, or come in at 9 AM, it’s safe to assume that someone will be in later or earlier than you to work on their research. Because of this, emphasizing a healthy work-life balance is important, as it can help lower stress levels and allow researchers to better appreciate their lives holistically.

Getting Gases Colder

Back inside the laboratory, things were quickly shifting. Wieman left JILA "in 2006 to found a physics education research program at the University of British Colombia," said *JILA, the First 50 Years*. To try to fill the hole in ultracold research that Wieman had left behind, Debbie decided to collaborate with JILA Fellow and colleague Jun Ye, a decision that would prove to be immensely successful.

Ye was already quite familiar with JILA by the time he and Debbie started collaborating in the early 2000s. Ye had originally gone to JILA for his PhD., studying under John "Jan" Hall. Like Imes in chapter 3, Ye's Ph.D. thesis focused on molecular spectroscopy, though obviously using more modern equipment than Imes had back in the early 1900s. After receiving his degree in 1997, Ye went to the California Institute of Technology for his postdoc in Jeff Kimble's quantum optics group. He returned to JILA in 1999 as an associate Fellow and a NIST physicist. Upon Ye's arrival, Jan Hall donated most of his laboratory space to Ye. In 2001, Ye became a full JILA Fellow. Like Debbie, Ye was fascinated with ultra-cold molecules, though he also worked on optical atomic clocks, the most precise clocks in the world.

Ye originally didn't think about collaborating with Debbie but was encouraged by Wieman (before he left), who predicted a brilliant partnership in the making. Because the W.M. Keck Foundation, a philanthropic organization focused on science and engineering, hoped to donate $1.5 million to JILA for research in ultracold molecules, Wieman saw this donation as an opportunity to start a specific ultracold research group within JILA. The group would consist of himself, Cornell, Ye, and Debbie. As Ye recalled "[Carl] said, you two are the youngest Fellows [at the time], I have lots of expectations for you and Debbie, and I want to get you two together." Ye felt honored to be added to the group, as Debbie was already a well-known scientist. "For me was like, 'what?' That was kind of a dream," he said to Kenna. "I was very interested in cold atoms. And I learned some tricks at Caltech when I was a postdoc." With Ye's background in lasers and Debbie's in magnetic traps, the combination of the two would prove to be extremely successful.

It did however take a while to get there. "It was a very difficult experiment," said Ye. "I would say that even when we first started, the vision wasn't so clear." Not sure what to expect, Ye and Debbie plowed ahead. By 2006, their laboratory groups were meeting together and they were often working together in the lab. Even with all this hard work, the first ground-breaking results would only show up five years after the cold-molecule center officially started. In 2008 "Jin and Ye succeeded in crafting tens of thousands of ultracold polar potassium-rubidium (KRb) molecules in their lowest energy state, creating an entirely new form of quantum matter in the process," explained *JILA: the First 50 Years*. This was an important step in advancing ultracold quantum research, as it helped open the door to further projects in ultracold chemistry. The research also gave insights into chemical reactions at temperatures near zero Kelvin. "We showed how to use lasers to transfer molecules to the absolute ground state," explained Ye. "And that really opened the field up. The molecules are fermions, and we can change their characters (being identical fermions or not) by simply switching their nuclear spin states. At that time, I witnessed, as a scientist, the wonderfulness of Debbie. From Debbie, I learned how you pay attention to details not only in the lab but also in data analysis and in theoretical modeling." Debbie also taught Ye how to better separate work and home life. "Another thing that I learned from Debbie was that when there are thousands of things pulling you apart, especially emails, she just turned it off. She just didn't answer any emails. That was actually very educational to watch her move into a science mode."

While achieving this system not only enabled further research, it could help with "designer chemistry," a 2016 NIST article stated. Because quantum physics can explain interactions on a subatomic level, it offers huge potential benefits for chemistry, especially in drug design and development. These gases can also model more complex systems, including those inside a quantum computer. "This [ultracold gas] is one architecture you could use for a quantum computer, where these cold atoms would each serve as your quantum bit; where each one could store a quantum bit of information," Debbie stated in her 2014 lecture for receiving the Institute of Physics (IOP) Isaac Newton Medal and Flag. "This is only one of a number of possibilities for quantum computing, but it is an architecture for quantum computing that is being actively pursued." While Debbie didn't directly look into a quantum computing connection to her work, many of her colleagues at JILA continue to use some of her methods for applications in quantum communication and quantum information.

In 2011, Debbie, Ye, and their teams were able to improve their experiment by changing the spatial dimensions of the gas of the molecules. Unlike atoms, molecules can interact with each other over large distances because they carry something called an electric dipole moment. Furthermore, how they interact with each other depends on the orientation of the molecules that can be controlled with laboratory fields. This really opened up new ways to see new phenomena mediated by interactions in the quantum regime. Once JILA added its new X-Wing to the building in 2012, the researchers were able to move their experiments into a much larger laboratory. From there, Ye and Debbie proved to be a dynamic duo within the scientific community. As Ye said "People wanted to see whether that model [a pair of scientists] can be replicated in other places. There were many people seeing that we had this great pairing and say 'they're solving some problems that couldn't be solved before.' We felt really honored. I mean, for me, in particular, because I feel like Debbie was such an amazing scientist that even to be mentioned in the same breath is a great honor."

Working together on ultracold molecules created a deep friendship between Debbie and Ye. 'We had regular dinners at each other's homes," added Ye. "She and I traveled to quite a few places for conferences." Ye's children also grew up with Jackie, bringing both families closer together as well. Ye saw Debbie not only as a friend but also as a mentor. There was one incident that Ye writes about in the 2017 NIST article that showed Debbie's confidence in both her and Ye's skills as researchers:

> "I had come to respect Debbie so much that I started asking her for advice on matters outside of science. I will never forget one instance. I had a number of universities that approached me for possible relocation. I did not really consider them seriously. But one time, an institution with several colleagues whom I really respect approached me. I went to Debbie's office and said, 'Yesterday I received calls from these colleagues…' Debbie smiled at me, with her deep dimples, and said calmly, 'Well Jun, you like working with me, right?' And I said 'Yes…' 'But I am not moving,' Debbie said."

Recalling that memory, Ye said to Kenna: "I feel that some young people analyze problems in a complex way. You focus on too many things, like 'how much money I'm getting, how much funding I'm getting, etc.' Debbie was able to distill into a single point just: 'here's my colleague, and do I enjoy this colleague?' And if you do, and if the colleague is not moving, then you shouldn't move." And Ye didn't, continuing to collaborate with Debbie until her death in 2016.

Debbie as a Mentor and Advocate

In the following years, Debbie and her students continued to study the dynamics within the ultracold realm, uncovering more insights into the world of quantum physics. She also made sure to be the best mentor that she could be to her students. "As an undergraduate, I did research in one of Debbie's labs," JILA alumn Travis Nicholson wrote on the virtual memorial wall. "Working for her was an amazing experience. She was a kind and generous advisor and she was committed to the idea that her students should receive first-rate scientific training. Debbie often took time out of her busy schedule to sit down with me and make sure I was learning important concepts and techniques, to make sure I was staying on top of my classes, and to provide mentorship for the next phase of my career. I always looked up to Debbie, and she remained a trusted friend after I graduated." Others echoed this sentiment, sharing stories on the memorial wall about Debbie's quickness to answer questions and explain difficult science concepts in a digestible way. “She was all at once a mentor, friend, and role model,” Regal wrote on the wall. “Her effect on my life has been immeasurable. I will remember her calm poise, her smile, her capacity to be utterly absorbed in a physics experiment, and her love of family. Scientifically she never overcomplicated any subject, and she knew how to find an exciting problem that was at the same time possible in the laboratory. She was a unique force and her wise advice will live on with the many students she devoted so much to. Regal told Kenna about some of the advice that Debbie gave her before Regal initially left JILA in 2006. One of the pieces of advice on teaching was “You don't have to understand everything, just learn faster than the students.”

As a leading scientist in her field, Debbie not only championed scientist-mothers but also women in science in general. “Physics was a little bit of a boy's treehouse club thing,” Cornell said in a 2016 article for the *Boulder Daily Camera.* “But when she burst into that club, she didn’t pull up the ladder afterward. Around the world, there are many people, many women, that she trained who are now professors in their own right at great universities. She was an inspiration to a generation of female scientists.” Debbie’s role model for women in science became even more solidified when she was awarded the 2013 L’Oréal-UNESCO For Women in Science award. The partnership between L’Oréal and UNESCO began in 1998 and has focused on recognizing women in science and supporting their research by selecting the top woman scientist from each continent each year. Debbie was cited by the awards committee for “having been the first to cool down molecules so much that she can observe chemical reactions in slow motion, which may help further understanding of molecular processes which are important for medicine or new energy sources.” Debbie’s driving spirit was the main force in her for advocating other women in science. “If every picture on the wall of a physicist is a man, you just start to have more self-doubt,” she said in her speech for the 2013 award. According to John, Debbie thought that her interactions with other brilliant women while in Paris to receive this award was a highlight of her career. While Debbie did promote women in physics through various groups and lectures, she also did her best to live by example. “She is still considered a prominent woman in science, and she still is absolutely a role model,” John added.

Later Years and Legacy

No doubt Debbie’s important work in both science and advocacy would have continued if she hadn’t been diagnosed with cancer. Being a private individual, Debbie told only a few people outside of her family about the diagnosis in early 2016. “She was in tears only once, when she initially shared with me in late January that she had cancer and that’s only because she was trying to comfort me when I started to cry,” Ye stated in the 2017 NIST article. DeMarco was also told early on. “I made sure to spend more time with her and made various excuses to visit,” DeMarco told Kenna tearfully. He, like Ye and many others, still finds it hard to discuss Debbie’s passing. “Not a lot of people knew what was happening until quite late,” DeMarco added. "Privacy was important to her.” Ye, knowing earlier on about Debbie’s diagnosis, transitioned into a more responsible role for their joint group of students working on cold molecules, understanding that Debbie wouldn’t be in the laboratory for a while. He had a photo taken of Debbie with a few students and himself in April 2016, the last time she would be in JILA. “You know at that point, I don't know what I was thinking, but I knew this could be her last time [in a while],” Ye said to Kenna. At this point, Ye and Debbie had already decided to rebuild the machine needed for seeing a new range of scientific phenomena with KRb molecules. Ye told Kenna it was “poetically ironic” that Debbie, who was sick and “withering,” was standing next to this machine that was being built from scratch. After Debbie stopped coming into the laboratory, Ye would visit her often and catch her up on the research, often drawing pictures of the apparatus and taking her for walks in the Colorado sunshine around her neighborhood. Cornell also made similar visits, brightening Debbie’s day and making her feel connected to her community at JILA.

For those who found out about Debbie’s diagnosis later in 2016, closer to the end, the news was devastating. Even during the uncertainty of her cancer treatment, Debbie was still hard at work on her research and supporting others. "Debbie Jin found time during her illness this year to mentor me through my own cancer treatment," Julie Phillips, a former JILA staff member, wrote on the memorial wall. "For a few months, we were girlfriends, but instead of talking about physics or our husbands, we talked about cancer and how best to deal with the demands of chemo and other treatments. We took lovely walks and just lived life. When I was scared about my third chemo infusion because of worsening neuropathy, Debbie came and sat with me through the whole thing. This was an astonishing act of kindness I will never forget."

Ye recalled to Kenna the last time he saw Debbie before her passing. His wife had cooked Bok Choi, one of Debbie's favorite foods. Ye shared one last meal with her, as Debbie tried to stay upbeat in her weakened state. In early September, Debbie arrived at the hospital where she would pass away on September 15, 2016, at the age of 47. Ye, Cornell, and John all played important roles at Debbie's funeral, and as Ye told Kenna, he felt incredibly heavy with grief. The community also had a "celebration of life" at Boulder's Millenium Hotel. As John told Kenna "people came from all over, and we overflowed a room designed to accommodate 200 people." The communities at JILA, NIST, CU Boulder, and beyond mourned her passing, and still continue to cherish her memories to this day.

"Jun, you have to take care of my students," Debbie told Ye at the very end. "They were still very important to her," Ye said tearfully. And he and Cornell did just that. They met with every single one of Debbie's students and fostered them, moving them into their own laboratories, and in some cases switching their research projects entirely. Both Ye and Cornell worked hard to try and be as good of mentors as Debbie was to her students. All the students that Debbie left behind are now successful scientists and academics, and some of them still look to her for inspiration in their own work.

Many in the scientific community believed, and continue to believe, that if it wasn't for her death, Debbie would have probably been awarded the Nobel Prize at some point due to her advances in quantum physics. "I've no doubt that she would have been a future Nobel laureate in due course, as one of the first experimenters in the unitary regime of fermions," JILA alumn Peter Drummond wrote on the memorial wall. Others agreed with this sentiment. One of Debbie's old colleagues, Kathryn J. Levin, added that Debbie "probably would have gotten the Nobel," in a 2020 article. As the regulations for the Nobel Prize clearly state that the prize cannot be awarded posthumously (unless the recipient was already announced and then passed away before receiving the award), many scientists like Debbie miss this opportunity to be recognized for their ground-breaking work. However, it probably wouldn't have really mattered to Debbie. As Ye said to Kenna, "she would have asked me 'what does actually change?'" While Debbie was not able to be awarded this prestigious prize, she was honored in many other ways. In 2002, *Discover* magazine named her as one of the 50 most important women in science.

To further honor her legacy, CU Boulder's College of Arts & Sciences created the "Deborah Jin Endowed Graduate Fellowships Fund." The fund, starting in 2020, would fully financially support the mentorship and research of two physics graduate students each semester. DAMOP also named their Award for Outstanding Doctoral Thesis Research in Atomic, Molecular, or Optical Physics after Debbie to remember her legacy. According to the DAMOP webpage: "DAMOP launched a supplemental campaign in 2016 to recognize Deborah Jin's contributions to physics by establishing the Deborah Jin Memorial Endowment. Upon achieving and exceeding the goal through the generosity of Deborah Jin's colleagues, family, and friends, DAMOP renamed the award in her honor. The income from this endowment will supplement travel for finalists and provide new funding to enable students to attend the annual DAMOP Meeting, with an emphasis on enhancing the diversity of student attendees." The University of Chicago, Debbie's alma mater, also created a fellowship in her honor. No doubt Debbie would have been pleased to see students being taken care of with these many different fellowships and awards.

Additionally, the Australian Research Council Center for Excellence for Engineered Quantum Systems also created the Deborah S. Jin fellowship. According to Dr. Liz Bridge, a recipient of this fellowship, "the fellowships were dedicated to supporting and promoting the work of early-to-mid career quantum researchers who are women, gender diverse, or identify as women, with the aim of going some way to addressing the under-representation of women in this field. I feel incredibly honored to have been awarded this fellowship in Debbie's namesake, and it had a massive impact on my career and self-confidence as a quantum researcher. For me, this is just one of many ways in which Debbie's legacy lives on and I will be forever thankful for the impact she's had on my career."

While all the students in Debbie's laboratory have now graduated and moved on, Ye, Cornell, and other JILA Fellows and scientists around the world continue with some of her research projects. In particular, Ye is still looking at ultracold KRb molecules, hoping to uncover more about the quantum realm, as the machine has now been fully rebuilt with new capabilities. "We finally got results in 2019 to 2022 and were able to publish five papers in *Nature* and *Science*," Ye said. "Debbie would be very proud of us. I actually printed out some of these papers and brought them to her tomb, and just laid them there for her to read."

Even for people who never met Debbie, she continues to play an important role in their lives. "Although I never had the pleasure of meeting Deborah Jin in person, she has had a huge impact on my career as a quantum physicist," Bridge said to Kenna. "Debbie was known across the globe as being a fantastic physicist and for her ground-breaking work with ultracold fermi gases and ultracold diatomic molecules. Working in the same field as Debbie, she has always been a role model for me, as one of the very few female physicists leading this field, and it was incredibly sad news to hear of Debbie's passing."

With her brilliance and dedication to the people of science as well as the science itself, Debbie stands out as a pillar in this community. "Debbie was so esteemed by her colleagues that everyone mentioned her work before her gender," explained JILA alumni Sara Campbell, "It was an important early example for me that a woman could be known first and foremost as a scientist." Her story continues to inspire others, and for this reason, it should be shared over and over. In researching and telling Debbie's story, both Kenna and Brian have gained a deeper appreciation for the hard work Debbie did to better her community. We hope that in sharing her story your own ideas of what a scientist should do or be are hopefully deepened a bit more. And perhaps, like many of us, she will play a role in your life too. “Debbie meant a lot to me,” Weidner stated, “She was one of the giants upon whose shoulders I stand. If I could be 1% of what she was, then I'd consider that a success.”

Conclusion

From reading these stories, it might be easy to think that these giants all were special in some unique way, which is somewhat true. However, you'll also notice that each of these individuals were just that, individuals. It was their work ethic, passion, and determinism that drove them to becoming giants, as well as their interest in supporting others. Those factors can be applied to nearly anyone, should they be interested in making similar impacts in their local communities. While saying that "anyone can be a giant" is a bit cliché, these stories illustrate this exact idea. A "giant" can be from anywhere in the world, from any time period.

Olena Mykhailova, Mariia Mykhailova's grandmother.

We see this happening right now with many of the talented individuals in our own lives. For Mariia Mykhailova, a Principal Software Engineer at Microsoft Quantum, the "giant" in her life turned out to be her grandmother, a female computer programmer working in Ukraine during the Ukrainian Soviet Socialist Republic (UkrSSR) since the 1960s

In her blog "My Granny is a Programmer," Mariia discusses how her grandmother, Olena Mykhailova, inspired both her own daughter (Mariia's mother) and Mariia to both pursue careers in computer science. As Olena went to university in the late 1940s, her class was the first to officially learn computer programming. As you've already read, those were the early days of computers, as Mariia wrote in her blog: "The curriculum included some brand-new theoretical courses: programming, design of analog and digital computers, computational methods (mostly methods of solving differential equations and linear algebra problems) etc. The lectures were attended not only by students, but also by employees of the Computational Center of the Academy of Science of the UkSSR- often there was no other source of this advanced knowledge."

Olena was one of the first women to join the Computational Center of the Academy of Science of the UkSSR in 1959. "The Center did work for companies which needed numerical solutions to some practical problems, the companies' representatives created a formal definition of a problem, and the Center's engineers solved it and coded the solution," Mariia explained in her blog. As Olena helped program the first computers used by the UkSSR she had many challenges to overcome. As Mariia illustrated: "Available machine time was always scarce, so programmers often stayed up late into the night to grab some extra time, while the punch card lab (a place where one could modify their punch cards using conventional machinery) closed at the end of the day. At night one had to edit their cards by hand - cut out the extra holes and paper over the excessive ones (using leftover pieces of card from conventional punching). The technicians disapproved of this technique - the holes cut extra could be tolerated, but the ones with pieces of card glued to them were dangerous - these pieces could fall off inside the input device and damage it." Olena later moved to other institutions to do programming there. As Mariia explained to Kenna, her grandmother's story inspired her to think about software engineering as a potential career. "When both your grandmother and your mother are software engineers, going into computer science yourself is a natural decision, since you have role models right there in your family," she stated.

Currently working at Microsoft Quantum, Mariia found her passion in joining quantum computing industry in its early days echoing similar work to what her grandmother did decades before by helping others learn quantum computing by teaching at a university and writing books. Olena's and Mariia's stories illustrate that the potential to be a "giant" can be found in nearly any individual, depending on their interests and dedication.

We also see from these stories that science is a community endeavor. Many of the "giants" in this book interacted with the same individuals (the biggest being Einstein) and all of them belonged to groups of fellow researchers or teams of workers. As Chris Ferrie writes in his book *Quantum Bullsh*t*: "Any particular field of science is a collection of thousands of people, each playing a small part in a solution that is built by a series of ever-improving failures." Because of this community aspect, all of our "giants" had help, support, and mentorship for others, and wouldn't have been able to succeed without these factors (in fact you could say that these "giants" had giants of their own). We can see that through history these community dynamics change due to war, prejudice, or other situations, but the community is never completely lost. And now, thanks to social media and the internet, these communities are more connected and vibrant than ever, suggesting that the next generation of "giants" for quantum computing could expand their influence and impact to even more groups.

In writing about these individuals, we shared their true natures. Nearly all were parents, nearly all dealt with death or loss at some point in their lives. And like all of us, they had to push through many obstacles to achieve their goals. That shouldn't sound so alien to us. We may think of "scientists" as a cold, distant breed, but hopefully this book helped widened the gaze of who belongs in this category. And as more and more stories like these pop up, the category will hopefully widen even further.

Many of these stories are being told for the first time, such as Debbie's or Mark's. This makes them even more important to share with others, as they will hopefully become adopted into a more public sphere. While this book highlights a few of these "unsung heroes" of quantum computing, mathematics, physics, and engineering, there are many other stories out there just like these, waiting to be shared. In our world of ever-expanding knowledge, wouldn't you agree that it's time to share them?

A Quick Favor Please?

Before you go may we ask you for a quick favor?
Good, we knew we could count on you.

Would you please leave this book a review on Amazon?

Reviews are very important for authors, as they help us sell more books. This will in turn enable me to write more books for you.

Please take a quick minute to go to Amazon and leave this book an honest review. We promise it doesn't take very long, but it can help this book reach more readers just like you.

Thank you for reading and thank you so much for being part of the journey.
-Brian & Kenna

Brian Lenahan is the Founder & Chair of the Quantum Strategy Institute. He is also the author of six Amazon-published books on artificial intelligence and quantum computing including the Bestseller "Quantum Boost – Using Quantum Computing to Supercharge Your Business". He is a former executive in a Top 10 North American bank, and mentors innovative companies in Canada areas. Brian's training in quantum computers comes from CERN/University of Oviedo, and Technische Universiteit Delft, and he writes extensively on quantum computing.

Leading multimillion-dollar programs, and teams of over 150 employees in the US, Canada and partnering with resources in India, Brian understands the demands of leadership. He has experience in numerous industries including Financial Services, Computer Services, Transportation, and Real Estate as well as functions including Strategy, IT, HR, Operations, Learning, and Financial Crimes & Fraud Management. He has consulted to C-Suite and Senior Management leaders for over 15 years.

Brian Lenahan is a recognized global keynote speaker, Global Quantum Top Voice for 2022 and 20323 and a developer of AI & QC strategies for large corporations and scaling firms. He is the CEO of Aquitaine Innovation Advisors (an AI & QC consulting firm) and the AI & Quantum Advisor at Haltech Regional Innovation Centre based in Burlington, Canada. Brian's experience skydiving, parasailing, zip lining canyons, and running half-marathons led to his personal approach of balancing prudent risk-taking while still seeking out commercial opportunities.

Brian is the author of four previous books on artificial intelligence, Amazon bestseller "Artificial Intelligence: Foundations for Business Leaders and Consultants", "Digital Coach: Coaching in the Era of Artificial Intelligence", and "Artificial Intelligence: Transitions – How to Successfully Prepare for Career Where AI is Everywhere", and "Deep Health: Using Artificial Intelligence to Live Longer and Healthier" all available on Amazon.com.

Aquitaine Innovation Advisors can be found at www.aquitaineinnovationadvisors.com

Brian Lenahan's posts are available on LinkedIn at: **https://www.linkedin.com/in/brian-lenahan-innovation/**

Kenna Hughes-Castleberry is a writer, podcaster, and science communicator. She currently works as the Science Communicator at JILA, and is the Editor-in-Chief of their journal *Light & Matter.* She is also a freelance science journalist, and writes for *Inside Quantum Technology* as a freelance staff editor. Her beats include deep technology, quantum technology, metaverse technology, and diversity within these industries. Kenna's work has been featured in various publications including *Scientific American, Discover Magazine, Physics.org, Inside Quantum Technology, The Quantum Insider, The Deep Tech Insider, the Metaverse Insider, The Debrief,* and *Octonation.* She currently sits on the board of SWARM (Science Writers Association of the Rocky Mountains) as well as teaches science writing to graduate students at JILA. Kenna has an M.Sc. in Science Communication from Imperial College London and lives in Broomfield, Colorado with her husband and her dog.

Find out more about Kenna at her website: https://kennacastleberry.com/

Follow Kenna at any of the following social media channels:

LinkedIn: https://www.linkedin.com/in/kenna-hughes-castleberry-m-sc-599a8b154/

Twitter: https://twitter.com/kennaculture

APPENDICES

BIBLIOGRAPHY

Chapter 1 – Al-Khwarizmi

Chandio, A. R. (2021). The house of wisdom (Bait Al-Hikmah): A sign of glorious period of Abbasids caliphate and development of science. In *International Journal of Engineering and Information Systems (IJEAIS) IS S* (Vol. 5). www.ijeais.org/ijeais

Chris Simms. (2020). Muhammad ibn Musa al-Khwārizmī-Persian polymath who created algebra. *New Scientist* .

DURAN, S., & SAMANCI, H. (2018). Al-Khwârizmî's Place and Importance in the History of Mathematics. *ITM Web of Conferences*, *22*, 01037. https://doi.org/10.1051/itmconf/20182201037

"Grover's Algorithm." Wikipedia. Wikimedia Foundation, December 2, 2022. https://en.wikipedia.org/wiki/Grover%27s_algorithm.

Lee, J., & Leung, L. (1998). *Mathematics in the Islamic World.* http://www-history.mcs.st-and.ac.uk/

Masud, I., & Syed, I. B. (2013). *Al-Khwārizmī-the Grand Father of Computer Science and Algebra.* Sunnah Muakada. https://sunnahmuakada.wordpress.com/2013/10/01/al-Khwārizmī-the-father-of-computer-science-and-algebra/

Nabirahni, D. M., Evans, B. R., & Persaud, A. (2019). Al-Khwarizmi (Algorithm) and the Development of Algebra. *Mathematics Teaching Research Journal.* https://commons.hostos.cuny.edu/mtrj/

"Quantum Cryptography - Shor's Algorithm Explained." RSS. Classiq Technologies, July 19, 2022. https://www.classiq.io/insights/shors-algorithm-explained.

Sesiano, J. (2008). Al-Khwarizmi. In *Encyclopaedia of the History of Science, Technology, and Medicine in Non-Western Cultures* (pp. 137–138). Springer Nature. http://www.ee.ryerson.

Shahid, I. (2018). Al-Khwarizmi: From Algebra to computing. *Australasian Muslim Times*.
Syed. Ibrahim B. (2006). *Al-Khwarizmi*. The History of Islam. https://historyofislam.com/contents/the-classical-period/al-khwarizmi/

van der Waerden, Bartel L. "A History of Algebra." A History of Algebra. Google, 2013. https://books.google.com/books?hl=en&lr=&id=W6DwCAAAQBAJ&oi=fnd&pg=PA3&dq=al%2Bkhwarizmi&ots=hLRNdBs078&sig=3vZXrbtkD1p6EnQY5mhzE04ZQ1k#v=onepage&q=al%20khwarizmi&f=false.

Waldman, Marilyn. "Conversion and Crystallization (634–870)." Conversion-and-crystallization-(634-870). Encyclopædia Britannica, inc., 2018. https://www.britannica.com/topic/Islamic-world/Conversion-and-crystallization-634-870.

Interviews:
Dr. David Nabrihni, professor at Pace University
Dr. Brian Evans, professor at Pace University
Dr. Keeper Sharkey, mathematician, and CEO of ODE and L3C
Dr. Scott Aaronson, professor at University of Texas at Austin
Chris Ferrie, senior lecturer at Centre for Engineer Quantum Systems of the University of Technology Sydney and author of *Baby University* books
Marina Walt, Product Manager of HQS Quantum Simulations

Chapter 2 – Leibniz

Ball, Philip. "Polynesian People Used Binary Numbers 600 Years Ago." Scientific American. Scientific American, December 16, 2013.

https://www.scientificamerican.com/article/polynesian-people-used-binary-numbers-600-years-ago/

Buchanan, M. (2019). Leibniz's principle of indistinguishability. In *Nature Physics* (Vol. 15, Issue 11, p. 1104). Nature Publishing Group. https://doi.org/10.1038/s41567-019-0711-5

Burnham, D. (2022). Leibniz, Gottfried_ Metaphysics _ Internet Encyclopedia of Philosophy. *Internet Encyclopedia of Philosophy*

Ghose, Tia. "Math Surprise: Remote Islanders Invented Binary Number System." LiveScience. Purch, October 14, 2022. https://www.livescience.com/41985-binary-math-invented-before-leibniz.html.

"Gottfried Wilhelm Leibniz." Leibniz Mathematical School, November 9, 2022. https://leibniz-math.org/gottfried-wilhelm-leibniz/.

History.com Editors. "Printing Press." History.com. A&E Television Networks, May 7, 2018. https://www.history.com/topics/inventions/printing-press#section_1

Ivancheva, L. (2021). *Leibniz's Monadology and its insights concerning quantum physics Ludmila Ivancheva.*

Kehoe, Joselle. "Quantum Mechanical Words and Mathematical Organisms." Scientific American Blog Network. Scientific American, May 16, 2013. https://blogs.scientificamerican.com/guest-blog/quantum-mechanical-words-and-mathematical-organisms/

"Merck Group." Wikipedia. Wikimedia Foundation, November 29, 2022. https://en.wikipedia.org/wiki/Merck_Group.

Schreiber, Barbara A. “De Broglie Wave.” Encyclopædia Britannica. Encyclopædia Britannica, inc. Accessed November 29, 2022. https://www.britannica.com/science/de-Broglie-wave

White, Matthew. “The Enlightenment.” British Library, 2018. https://www.bl.uk/restoration-18th-century-literature/articles/the-enlightenment

Wolfram, S. (2013). *Dropping In on Gottfried Leibniz—Stephen Wolfram Writings*.

Interviews:
Dr. Keeper Sharkey, mathematician, and CEO of ODE and L3C
Chris Ferrie, senior lecturer at Centre for Engineer Quantum Systems of the University of Technology Sydney and author of *Baby University* books
Dr. Thomas Ehmer, the Innovation Incubator Lead at Merck Group
Peter Bordow, quantum systems leader at Wells Fargo
Dr. Robert Kleinberg, professor of computer science at Cornell University

Chapter 3 – Imes

Admin. “Spectroscopy: Types of Spectroscopy: Infrared & Raman Spectroscopy.” BYJUS. BYJU'S, August 2, 2022. https://byjus.com/chemistry/spectroscopy/.
“Background: Introduction to spectroscopy20.” NASA. NASA, October 14, 2021. https://imagine.gsfc.nasa.gov/educators/lessons/xray_spectra/background-spectroscopy.html.

“Baltimore Afro-American.” Google News Archive Search. Google, October 21, 1923. https://news.google.com/newspapers?nid=KD63RgGzVVoC&dat=19331021&printsec=frontpage&hl=en.

Bennett, Kathy. "Lynching." Tennessee Encyclopedia. Tennessee Historical Society, March 1, 2018. https://tennesseeencyclopedia.net/entries/lynching/.
Bois, W. E. B. du. (1931). *Prepared by the Center for History of Physics at AIP 1 Physicist Activist: Dr. Elmer Imes and the Civil Rights Case of Juliette Derricotte Primary Source Letter from Elmer S. Imes to.* http://credo.library.umass.edu/view/pageturn/mums312-b189-i019/#page/1/mode/1uphttp://credo.library.umass.edu/view/full/mums312-b189-i019

Center for History of Physics. (2012). *CENTER FOR HISTORY OF PHYSICS AT AIP A Brilliant Scientist*. www.aip.org/history-programs.

Cox, T. (2020) Spectroscopy in Everyday Life. Chem ED Exchange. Available at: https://www.chemedx.org/activity/spectroscopy-everyday-life.

Davis, R. m. (2022). The Work of Elmer Imes. *Symmetry Magazine*.

"Elmer Imes." Wikipedia. Wikimedia Foundation, August 10, 2022. https://en.wikipedia.org/wiki/Elmer_Imes.

"Elmer Imes and Spectroscopy - American Institute of Physics." Dr. Elmer Imes and Spectroscopy_Elmer Imes Handout.pub, 2002. https://www.aip.org/sites/default/files/history/teaching-guides/imes-spectrometers/Dr.%20Elmer%20Imes%20and%20Spectroscopy_PowerPoint.pdf

"Emmett Till." The United States Department of Justice, September 12, 2022. https://www.justice.gov/crt/case/emmett-till-0.

Ferrie, Chris. *Quantum Bullshit: How to Ruin Your Life with Advice from Quantum Physics*. Naperville, IL: Sourcebooks, 2023.

Fikes, R. (2017). Elmer Samuel Imes (1883-1941) •. *Black Past.*
"Harrison M. Randall." Michigan Physics, April 3, 2015. https://michiganphysics.com/history/history-by-person/harrison-m-randall-2/.

Hutchinson, George. *In Search of Nella Larsen a Biography of the Color Line.* Cambridge, Mass: Belknap Press, 2006.

Imes, E. S., & Larsen, N. (2018). *Elmer Samuel Imes-Astrophysicist Background Led to Drive for Education And Became an Internationally-Recognized Physicist.* https://www.encyclopedia.com/people/literature-and-arts/american-literature-biographies/elmer-samuel-imes

Martin, Alfred E. "Elmer Samuel Imes, 1883–1941." American Association of Physics Teachers. American Association of Physics TeachersAAPT, January 1, 1970. https://aapt.scitation.org/doi/pdf/10.1119/1.1990348

Mickens, R. E. (1999). *The African American Presence in Physics.*

Mickens, R. E. (2018). The life and work of Elmer Samuel Imes. In *Physics Today* (Vol. 71, Issue 10, pp. 28–35). American Institute of Physics Inc. https://doi.org/10.1063/PT.3.4042

Swann, W. F. G. (1941). Elmer Samuel Imes. *Science*, *94*(2452), 600–601. https://doi.org/10.1126/science.94.2452.600

Interviews:
Robert Davis, Science Writer and Communications Specialist for IBM
Cedric Wilson, fifth-year PhD student in the Ultracold Quantum Gases Group within the MIT-Harvard Center for Ultracold Atoms (CUA).

Chapter 4 – Bose

Bose, S. N. (1924). Planck's Law and the Light Quantum Hypothesis. *Zeitschrift Fur Physik, 1.*

"Bose-Einstein: Osa History Gallery." Optica, 1953. https://www.optica.org/en-us/history/multimedia/photos/b/bose-einstein/.

"Bose–Einstein Statistics." Wikipedia. Wikimedia Foundation, December 14, 2022. https://en.wikipedia.org/wiki/Bose%E2%80%93Einstein_statistics.

Coffey, Tonya. "Bose Einstein Statistics." YouTube. YouTube, November 14, 2016. https://www.youtube.com/watch?v=mhSdPFlroDc.

Dance, Rozlynn, and Tessa Kaplan. "Encouraging Mathematical Confidence." Encouraging Mathematical Confidence, September 8, 2018. https://blog.heinemann.com/thinking-together-encouraging-mathematical-confidence.

Dasgupta, D. (2012). Creating a peripheral trading zone: Satyendra Nath Bose and bose-einstein statistics, doing science in the role of an outsider. *International Studies in the Philosophy of Science*, *26*(3), 259–287. https://doi.org/10.1080/02698595.2012.731731

Doyle, Bob. Doyle, Derek. Satyendra Nath Bose, 2016. https://www.informationphilosopher.com/solutions/scientists/bose/.

Ghose, Partha. "Counting in the Dark." Inference. Inference, October 13, 2022. https://inference-review.com/article/counting-in-the-dark.

Gregersen, Erik. "Satyendra Nath Bose." Encyclopædia Britannica. Encyclopædia Britannica, inc., 2022. https://www.britannica.com/biography/Satyendra-Nath-Bose.

"How Did Imperialism in India Come about?" CliffsNotes. Course Hero, 2022. https://www.cliffsnotes.com/cliffsnotes/subjects/history/how-did-imperialism-in-india-come-about#:~:text=The%20British%20took%20control%20of,known%20as%20the%20Sepoy%20Mutiny.

K, Diana. "Top 10 Unbelievable Facts about Satyendra Nath Bose." Discover Walks Blog, May 8, 2022. https://www.discoverwalks.com/blog/india/top-10-unbelievable-facts-about-satyendra-nath-bose/.

Mehra, J. (1975). Satyendra Nath Bose. *Biographical Memoirs of Fellows of the Royal Society*, *21*(21).

Mukhopadhyay, Mayukh. "SN Bose - the Ignored Genius." Medium. Medium, January 2, 2021. https://mayukhdifferent.medium.com/sn-bose-the-ingorned-genius-addcb541b058.

Nauenberg, M. (2007). Stories and statistics of bose. *Physics Today*, *60*(6), 12–14. https://doi.org/10.1063/1.4796447

Oliveboard. "List of Viceroys of India: Tenure and Incidents during Their Regime." Oliveboard, April 29, 2022. https://www.oliveboard.in/blog/viceroy-of-india/.

Petroff, Alanna. "The Exact Age When Girls Lose Interest in Science and Math." CNNMoney. Cable News Network, February 28, 2017. https://money.cnn.com/2017/02/28/technology/girls-math-science-engineering/index.html.

Sack, Harald. "Satyendra Nath Bose and the Einstein-Bose Statistics." SciHi Blog, January 1, 2017. http://scihi.org/satyendra-nath-bose-einstein-statistics/.

"Satyendra Bose." Physics Today. American Institute of Physics, January 11, 2017. https://physicstoday.scitation.org/do/10.1063/PT.5.031387/full/.

"Satyendra Nath Bose." Satyendra-Nath-Bose:: University of Dhaka, 2022. https://du.ac.bd/notablealhumnidetails/Satyendra-Nath-Bose/156.

"Satyendra Nath Bose." Wikipedia. Wikimedia Foundation, December 13, 2022. https://en.wikipedia.org/wiki/Satyendra_Nath_Bose.

Subramanian, Kapil. "Remembering the Classical Academician's Spirit of Satyendra Nath Bose." The Wire Science, June 4, 2022. https://science.thewire.in/the-sciences/remembering-classical-academicians-spirit-satyendra-nath-bose/.

"Unified Field Theory." Wikipedia. Wikimedia Foundation, September 15, 2022. https://en.wikipedia.org/wiki/Unified_field_theory.

Wali, Kameshwar C. "Satyendra Nath Bose -- His Life and Times: Selected Works (with Commentary) ." Google Play Books. World Scientific Publishing, 2009. https://play.google.com/books/reader?id=B2nICgAAQBAJ&pg=GBS.PR37&hl=en.

Wolpert, Stanley A. "British Raj." Encyclopædia Britannica. Encyclopædia Britannica, inc., September 2022. https://www.britannica.com/event/British-raj.

Interviews:
Dr. Scott Aaronson, Professor at the University of Texas at Austin
Mayukh Mukhopadhyay, data scientist
Santanu Ganguly, expert in quantum networking for Photonic Inc.
Abhijit Roy, software engineer in Toronto

Chapter 5 – Tomonaga

Adler, Jerry. "When Albert Einstein Visited Japan." Smithsonian.com. Smithsonian Institution, May 1, 2018. https://www.smithsonianmag.com/history/albert-einstein-visited-japan-180968737/

Ahmed, Samira. "How the Bomb Changed Everything." BBC Culture. BBC, July 2, 2015. https://www.bbc.com/culture/article/20150702-how-the-bomb-changed-everything

Atiyah, M. "Science for Evil: The Scientist's Dilemma." BMJ (Clinical research ed.). U.S. National Library of Medicine, August 14, 1999. https://www.ncbi.nlm.nih.gov/pmc/articles/PMC1127049/

"Dr. Tomonaga Sin-Itiro." Dr. TOMONAGA Sin-Itiro - University of Tsukuba, 2022. https://www.tsukuba.ac.jp/en/about/history/nobel/tomonaga/

Estate of Richard Feynman. "Richard Feynman." The Official Site of Richard Feynman, 2021. http://www.richardfeynman.com/about/bio.html

Francis, Charles. Hoath, Mike. and Toth, Victor T. "How Did Paul Dirac Unify Quantum Mechanics and Special Relativity? - Quora," 2016. https://www.quora.com/How-did-Paul-Dirac-unify-quantum-mechanics-and-special-relativity.
Fukuda, N., Miyamoto, Y., & Miyazima, T. (1966). To Sin-itiro Tomonaga on his Sixtieth Birthday. *The Progress of Theoretical Physics*.

Hayakawa, S. (1991). Sin-Itiro Tomonaga and His Contributions to Quantum Electrodynamics and High Energy Physics. *Progress of Theoretical Physics Supplement, 105*. https://doi.org/10.1143/PTPS.105.157/1837398

Jivkova, Kat. "Paul Dirac and the Origins of Quantum Mechanics." Retrospect Journal, November 15, 2020. https://retrospectjournal.com/2020/11/08/paul-dirac-and-the-origins-of-quantum-mechanics/.

Johannes, Henn., Chaniotis, Angelos., Fassin, Didier., and Voevodsky. Vladimir. "Mission & History." Institute for

Advanced Study, December 13, 2022.
https://www.ias.edu/about/mission-history
.

Konuma, M., Bando, M., Gotoh, H., Hayakawa, H., Hirata, K., Ito, K., Ito, K., Kanaya, K., Konagaya, D., Kugo, T., Namba, C., Nishitani, T., Takaiwa, Y., Tanabashi, M., Tanaka, K., Tanaka, S., Ukegawa, F., & Yoshikawa, T. (2014). *The Legacy of Hideki Yukawa, Sin-itiro Tomonaga, and Shoichi Sakata: Some Aspects from their Archives*.

Krieger, David. "The Russell-Einstein Manifesto." Nuclear Age Peace Foundation, April 5, 2018.
https://www.wagingpeace.org/russell-einstein-manifesto/

"Lamb Shift." Wikipedia. Wikimedia Foundation, August 1, 2022. https://en.wikipedia.org/wiki/Lamb_shift

Li, Wei. "The Mental Health Crisis in Science." Science in the News, March 5, 2021.
https://sitn.hms.harvard.edu/flash/2021/the-mental-health-crisis-in-science/

Low, M. (2005). *Science and the Building of a New Japan*. John Hopkins University.
www.columbia.edu/cu/weai/publications.html

Matsui, M., & Ezawa, H. (1995). Sin-itiro Tomonaga: The Life of a Japanese Physicist. *Scientific Publishing Division of MYU*.
Matsui, M., & Schweber, S. S. (1996). Sin-itiro Tomonaga: Life of a Japanese Physicist. *Physics Today*, *49*(7), 59–60.
https://doi.org/10.1063/1.2807690

Nave, R. "The Lamb Shift." The lamb shift. Georgia State University, 2016.
http://hyperphysics.gsu.edu/hbase/quantum/lamb.html

"The Nobel Prize in Physics 1932-Werner Heisenberg." NobelPrize.org, 2022.
https://www.nobelprize.org/prizes/physics/1932/summary/

"The Nobel Prize in Physics 1933-Paul Dirac." NobelPrize.org, 2022. https://www.nobelprize.org/prizes/physics/1933/dirac/biographical/

"The Nobel Prize in Physics 1945-Wolfgang Pauli." NobelPrize.org, 2022. https://www.nobelprize.org/prizes/physics/1945/pauli/biographical/.

"The Nobel Prize in Physics 1965." NobelPrize.org, 2022. https://www.nobelprize.org/prizes/physics/1965/tomonaga/lecture/

"Renormalization Theory." RIKEN, 2022. https://www.riken.jp/en/about/history/story/renormalization/.

Rouze, Michael. "J. Robert Oppenheimer." Encyclopædia Britannica. Encyclopædia Britannica, inc., 2022. https://www.britannica.com/biography/J-Robert-Oppenheimer.

Schweber, Silvan S. "Sin-Itiro Tomonaga." Essay. In *QED and the Men Who Made It: Dyson, Feynman, Schwinger, and Tomonaga*. Princeton, , NJ: Princeton Univ. Press, 2011.

Senami, M. (2019). Quantum Electrodynamics and Molecules. In *Reference Module in Chemistry, Molecular Sciences and Chemical Engineering*. Elsevier. https://doi.org/10.1016/b978-0-12-409547-2.11487-8

"Sin-Itiro Tomonaga." MagLab, 2012. https://nationalmaglab.org/education/magnet-academy/history-of-electricity-magnetism/pioneers/sin-itiro-tomonaga.

"Shin'Ichiro Tomonaga." Physics Today. American Institute of Physics, March 31, 2015. https://physicstoday.scitation.org/do/10.1063/pt.5.030932/full/

Shinichiro Tomanaga papers. "SHIN'ICHIRŌ TOMONAGA." Tomonaga, Shin'ichirō, 1906-1979, 2021. https://history.aip.org/phn/11811005.html

"Shin'ichirō Tomonaga." Wikipedia. Wikimedia Foundation, October 5, 2022. https://en.wikipedia.org/wiki/Shin%27ichir%C5%8D_Tomonaga
.

Tomonaga, S. (1966). Development of Quantum Electrodynamics: Personal Recollections. *Science*, *154*.
Wood, Charlie. "How Mathematical 'Hocus-Pocus' Saved Particle Physics." Quanta Magazine, February 26, 2021. https://www.quantamagazine.org/how-renormalization-saved-particle-physics-20200917/

Yamazaki, Toshimitsu. "Interplay between Yukawa and Tomonaga in the Birth of Mesons." arXiv.org, March 19, 2008. https://arxiv.org/abs/0712.4355

"Yoshio Nishina." Nuclear Museum, 2022. https://ahf.nuclearmuseum.org/ahf/profile/yoshio-nishina/.

Interviews:
Dr. Yidong Chong, professor at Nanyang Technological University

Chapter 6 – Shannon

Admin. "Boolean Algebra (Boolean Expression, Rules, Theorems and Examples)." BYJUS. BYJU'S, November 23, 2022. https://byjus.com/maths/boolean-algebra/.

Collins, G. P. (2002). Claude E. Shannon: Founder of Information Theory. Scientific American.

"Claude Shannon." Wikipedia. Wikimedia Foundation, December 15, 2022. https://en.wikipedia.org/wiki/Claude_Shannon.

"Claude E. Shannon: A Goliath Among Giants." Claude Shannon, 2022. https://www.belllabs.com/claude-shannon/.

"Claude E. Shannon: IEEE Information Theory Society." Claude E. Shannon | IEEE Information Theory Society, 2021. https://www.itsoc.org/about/shannon.

"Double Majoring: Is Two Always Better than One." Bay Atlantic University - Washington, D.C. Bay Atlantic University - Washington, D.C., June 14, 2022. https://bau.edu/blog/should-i-doublemajor/#:~:text=According%20to%20statistics%20reported%20by,%2C%20rising%20to%20around%2040%25.

Effros, M., & Poor, H. V. (2017). Claude Shannon: His Work and its Legacy. *EMS Newsletter*.

"Erector Set." Wikipedia. Wikimedia Foundation, September 17, 2022. https://en.wikipedia.org/wiki/Erector_Set.

Firor, John, and Virginia Trimble. "Thornton Carl Fry (1892–1991)." Bulletin of the AAS. PubPub, January 1, 1997. https://baas.aas.org/pub/thornton-carl-fry-1892-1991/release/1.

Goodman, Rob, and Jimmy Soni. "Betty Shannon, Unsung Mathematical Genius." Scientific American Blog Network. Scientific American, July 24, 2017. https://blogs.scientificamerican.com/voices/betty-shannon-unsung-mathematical-genius/.

"History." Nokia Bell Labs. Nokia Bell Labs, 2021. https://www.bell-labs.com/about/history/.

"Mars Home." MARS Home, 2016. http://mars.gmu.edu/.

Meye, Michal. "The Rise and Fall of Vannevar Bush." Science History Institute. Science History Institute, June 9, 2021. https://www.sciencehistory.org/distillations/the-rise-and-fall-of-vannevar-bush.

MIT Professor Claude Shannon dies; was founder of digital communications. (2001). MIT News. https://news.mit.edu/2001/shannonNoll

Memories: A Personal History of Bell Telephone Laboratories. A. M. (2015).

O'Connor, JJ, and E F Robertson. "Claude E Shannon - Biography." Maths History, October 2003. https://mathshistory.st-andrews.ac.uk/Biographies/Shannon/.

Parks, Abby. "Claude Shannon - Complete Biography, History and Inventions." History, December 8, 2022. https://history-computer.com/claude-shannon-complete-biography/.

Price, R. (1982). Oral-History:Claude E. Shannon - Engineering and Technology History.

"Remembering Claude Shannon." Survey. MARS GMU, 2002. http://mars.gmu.edu/jspui/bitstream/handle/1920/2051/index.html.1.2.html.

Roberts, S. (2016). Claude Shannon, the Father of the Information Age, Turns 1100100. The New Yorker.

Sarwate, Anand D. "Mary Elizabeth Moore Shannon, 95: IEEE Information Theory Society." Mary Elizabeth Moore Shannon, 95 | IEEE Information Theory Society, May 16, 2017. https://www.itsoc.org/news-events/recent-news/betty-moore-shannon-obituary.

Shannon, C. (1948). From The Mathematical Theory of Communication.

Shannon, C. E. (1937). SYMBOLIC AYSIS OF RELAY AND SWITCHING CIRCUITS.

Shannon, C. E. (1945). A Mathematical Theory of Cryptography.

Shannon, C. E. (1954). Universal Turing Machine with Two Internal States.

Soni, J. (2020). A Man in a Hurry: Claude Shannon's New York Years. https://jimmysoni.com/claude-shannons-new-york-years/

Soni, Jimmy, and Rob Goodman. "AMind at Play." Amazon. Simon and Schuster, 2019. https://www.amazon.com/Mind-Play-Shannon-Invented-Information/dp/1476766681.

"A Symbolic Analysis of Relay and Switching Circuits." Wikipedia. Wikimedia Foundation, April 25, 2022. https://en.wikipedia.org/wiki/A_Symbolic_Analysis_of_Relay_and_Switching_Circuits.

Tse, D. (2020). How Claude Shannon Invented the Future. Quanta Magazine. https://www.quantamagazine.org/print

Interviews:
Dr. Chris Ferrie, senior lecturer at Centre for Engineer Quantum Systems of the University of Technology Sydney and author of *Baby University* books
Dr. David Thomson of Queens University

Chapter 7 – Holberton

Alba, Michael. "Vacuum Tubes: The World before Transistors." Engineering.com, January 18, 2018. https://www.engineering.com/story/vacuum-tubes-the-world-before-transistors

American Institute of Physics. "Oral History of Herman Goldstine." Oral History Interviews | Herman Goldstine | American Institute of Physics, April 26, 2022. https://www.aip.org/history-programs/niels-bohr-library/oral-histories/4636

APG News. "Women's History Month: Forgotten Female Programmers of APG." APG News, April 1, 2020. https://apgnews.com/special-focus/apg-100/forgotten-female-programmers/

"Awards." ENIAC Programmers Project, July 19, 2022. https://eniacprogrammers.org/eniac-programmers-project/awards/#:~:text=ENIAC%20Programmer%20Betty%20Holberton%20received,the%20first%20ideas%20about%20compilation.%E2%80%9D

Brockmeier, Erica K. "75th Anniversary of the Electronic Numerical Integrator and Computer (ENIAC)." Almanac, February 11, 2021. https://almanac.upenn.edu/articles/75th-anniversary-of-the-electronic-numerical-integrator-and-computer-eniac

Brockmeier, Erica K. "The World's First General Purpose Computer Turns 75." Penn Today, December 6, 1970. https://penntoday.upenn.edu/news/worlds-first-general-purpose-computer-turns-75

Cameron, Lori, and Michael Martinez. "The Epic Journey of Women in Computing: Advances, Setbacks, and an Uncertain Future." The epic journey of women in computing. IEEE Computer Society, 2018. https://www.computer.org/publications/tech-news/research/history-of-women-programmers-in-computing-gender

Computer History: 1946 ENIAC Computer History Remastered FULL VERSION First Electronic Computer. YouTube. YouTube, 2015. https://www.youtube.com/watch?v=bGk9W65vXNA

da Cruz, Frank. "Programming the ENIAC." ENIAC programmers, 2001. http://www.columbia.edu/cu/computinghistory/eniac.html

Duffin, Erin. "Percentage of the U.S. Population with a College Degree by Gender 1940-2021." Statista, July 27, 2022. https://www.statista.com/statistics/184272/educational-attainment-of-college-diploma-or-higher-by-gender/

"Farm Journal Magazine | AgWeb." Farm Journal Magazine, 2022. https://www.agweb.com/farm-journal-magazine.
Fritz, W.B. "The women of ENIAC," in *IEEE Annals of the History of Computing*, vol. 18, no. 3, pp. 13-28, Fall 1996, doi: 10.1109/85.511940.

Gürer, D. (1998). *Women Role Models in Computer Science History*. https://archive.cra.org/CRN/html/9809/expanding/dg.2_2.t.shtml

"History." Electrical and Systems Engineering, 2022. https://www.ese.upenn.edu/history/

Hood, Kate, and Gertrude Beal. "Glossary-Birthright/Convinced Friend." Quaker.org, June 15, 2020. https://quaker.org/glossary/.

"J. Presper Eckert." Wikipedia. Wikimedia Foundation, December 2, 2022. https://en.wikipedia.org/wiki/J._Presper_Eckert

Janeczko, P. B. (2019). Introduction. *Secret Soldiers: How the U.S. Twenty-Third Special Troops Fooled the Nazis.*

"John Mauchly." Wikipedia. Wikimedia Foundation, December 2, 2022. https://en.wikipedia.org/wiki/John_Mauchly

Johnson, L.R. "Coming to grips with Univac," in *IEEE Annals of the History of Computing*, vol. 28, no. 2, pp. 32-42, April-June 2006, doi: 10.1109/MAHC.2006.27

"Kathy Kleiman." American University Washington College of Law, 2022. https://www.wcl.american.edu/impact/initiatives-programs/techlaw/our-team/Kathy-Kleiman/

Kleiman, Kathy. "Documentary Info." ENIAC Programmers Project, July 19, 2022. https://eniacprogrammers.org/documentary-info/

Kleiman, Kathy. *Proving Ground: The Untold Story of the Six Women Who Programmed the World's First Modern Computer.* New York, NY: Grand Central Publishing, Hachette Book Group, 2022.

Koerner, Brendan I. "How the World's First Computer Was Rescued from the Scrap Heap." Wired. Conde Nast, November 25, 2014. https://www.wired.com/2014/11/eniac-unearthed/

Koss, Adele M. "Programming on the Univac 1: a woman's account," in *IEEE Annals of the History of Computing*, vol. 25, no. 1, pp. 48-59, Jan.-March 2003, doi: 10.1109/MAHC.2003.1179879.

Kovo, Yael. "Kathy Kleiman: 'the Computers," The Remarkable Untold Story of the Eniac Programmers." NASA. NASA, September 3, 2018. https://www.nasa.gov/ames/ocs/seminars/kathy-kleiman

Levy, C. (2001). *Betty Holberton Dies.*

Levy, Steven "The Brief History of the ENIAC Computer." Smithsonian.com. Smithsonian Institution, November 1, 2013. https://www.smithsonianmag.com/history/the-brief-history-of-the-eniac-computer-3889120/

Lide, D. R. (2001). *NIST Special Publication #958: "A Century of Excellence in Measurements, Standards, and Technology."*
Little, Becky. "When Computer Coding Was a 'Woman's' Job." History.com. A&E Television Networks, September 1, 2017. https://www.history.com/news/coding-used-to-be-a-womans-job-so-it-was-paid-less-and-undervalued

Loff, Sarah. "When the Computer Wore a Skirt: Langley's Computers, 1935–1970." NASA. NASA, November 25, 2016. https://www.nasa.gov/feature/when-the-computer-wore-a-skirt-langley-s-computers-1935-1970

Meissner, L. (2016). *The Fortran Story Retold*

Nelson, E. C., & Metropolis, N. (1982). Early Computing at Los Alamos. *IEEE Annals of the History of Computing*, *4*(4).

"Our History." Central High School, 2020. https://centralhs.philasd.org/about-central-high-school/about-us/

Petzinger Jr. , Thomas. "History of Software Begins With Brainy Women's Work." The Wall Street Journal. Dow Jones & Company, November 15, 1996. https://www.wsj.com/articles/SB848012407846877000

"Pine Camp, Now Fort Drum, in the 1930s and 40s." North Country At Work. North Country Public Radio, June 26, 2018. https://www.northcountryatwork.org/collections/pine-camp-now-fort-drum-in-the-1930s-and-40s/

"Room Mother Definition & Meaning." Dictionary.com. Dictionary.com, 2022. https://www.dictionary.com/browse/room-mother

Ross, J. (1983). *Oral History Interview with Frances E. Holberton.*

Snyder, M. B. (1898). *August 1898 Harvard Astrophysical Conference.*

Snyder, William. "Monroe Benjamin Snyder." AncientFaces. AncientFaces, June 1, 2021. https://www.ancientfaces.com/person/monroe-benjamin-snyder-birth-1848-death-1932/192810421#memory-411701

Stern, N. (1977). *An Interview with J. Presper Eckert.*
Strasser, Michael. "Fort Drum Exhibit to Highlight Sonic Deception Training at Pine Camp during WWII." DVIDS. Defense Visual Information Distribution Service, 2020. https://www.dvidshub.net/news/382207/fort-drum-exhibit-highlight-sonic-deception-training-pine-camp-during-wwii

Tynker. (2020). *The ENIAC Programmers_ 'Refrigerator Ladies' No More - Tynker Blog.*

"UNIVAC LARC." Wikipedia. Wikimedia Foundation, April 27, 2022. https://en.wikipedia.org/wiki/UNIVAC_LARC.

Vogel, W. F. (2017). "The Spitting Image of a Woman Programmer": Changing Portrayals of Women in the American Computing Industry, 1958-1985. *IEEE Annals of the History of Computing.*

The Washington Post. "Frances Snyder Holberton, 84." Chicago Tribune. Chicago Tribune, December 11, 2001. https://www.chicagotribune.com/news/ct-xpm-2001-12-12-0112120287-story.html

Webster, Ian. "Inflation Rate between 1946-2022: Inflation Calculator." $96,200 in 1946 → 2022 | Inflation Calculator, 2022. https://www.in2013dollars.com/us/inflation/1946?amount=96200

"What Do Quakers Believe?" Quaker.org. Friends Publishing Company, August 15, 2022. https://quaker.org/faith-and-practice/.

"What Is a Vacuum Tube?" What is a Vacuum Tube?, March 13, 2021. https://www.computerhope.com/jargon/v/vacuumtu.htm#:~:text=Alternatively%20referred%20to%20as%20an,a%20switch%20or%20an%20amplifier

Winegrad, D. (1996). Celebrating the Birth of Modern Computing: The Fiftieth Anniversary of a Discovery at the Moore School of Engineering at the University of Pennsylvania. *IEEE Annals of the History of Computing* , *18*(1)

Interviews:
Phil Holberton, nephew of Betty Holberton

Priscilla Holberton, daughter of Betty Holberton
Elisa Torres Durney, leader of "Girls in Quantum"
Lorraine Tsitsi Majiri, President of OneQuantum Africa

Chapter 8 – Silaban

Davies, Sharyn. "Indonesia: The Rise of Moral Conservatism." Asia Media Centre | New Zealand, March 10, 2020. https://www.asiamediacentre.org.nz/opinion-and-analysis/indonesia-the-rise-of-moral-conservatism/

Goldberg, J. N., & Schucking, E. L. (2003). Peter Gabriel Bergmann. *Physics Today*, *56*(8), 64–66. https://doi.org/10.1063/1.1611361

Goldberg, J. N., & Silaban, P. (1976). Equations of Motion in the Null Formalism l'z. In *General Relativity and Gravitation* (Vol. 7, Issue 9).

"Grand Unified Theory." Force, Particles, Quantum, and Gravity - JRank Articles, 2022. https://science.jrank.org/pages/3095/Grand-Unified-Theory.html

"Grand Unified Theory." Wikipedia. Wikimedia Foundation, December 2, 2022. https://en.wikipedia.org/wiki/Grand_Unified_Theory#Motivation
.

"Joshua N. Goldberg." Wikipedia. Wikimedia Foundation, August 27, 2022. https://en.wikipedia.org/wiki/Joshua_N._Goldberg

"Konsistensi Pantur Silaban Dan Teori Fisika." viva.co.id, August 13, 2009. https://www.viva.co.id/arsip/82493-konsistensi-pantur-silaban-dan-teori-fisika

Krauss, Lawrence M. "A Brief History of the Grand Unified Theory of Physics." Nautilus, March 10, 2017. https://nautil.us/a-brief-history-of-the-grand-unified-theory-of-physics-236493/

Laksana, Ben K.C. "How the Daily Lives of Indonesian Youth Can Tell Us Why They Become More Conservative." The Jakarta Post. Accessed December 29, 2022. https://www.thejakartapost.com/academia/2020/03/03/how-the-daily-lives-of-indonesian-youth-can-tell-us-why-they-become-more-conservative.html

O'Brien, John. "Syracuse's Role in Historic Research of Einstein, Gravity Waves." Syracuse.com, February 15, 2016. https://www.syracuse.com/su-news/2016/02/syracuse_university_gravitatio.html

"Pantur Silaban." Wikipedia. Wikimedia Foundation, August 3, 2022. https://en.wikipedia.org/wiki/Pantur_Silaban

Permana, Adi. "ITB Contributes to Prepare Indonesia in Entering Quantum Era." Institut Teknologi Bandung, April 8, 2022. https://itb.ac.id/news/read/58553/home/itb-contributes-to-prepare-indonesia-in-entering-quantum-era

Permana, Adi. "Mengenang Prof. Pantur Silaban, Bapak Ilmu Relativitas Einstein Dari ITB." Institut Teknologi Bandung, August 4, 2022. https://www.itb.ac.id/news/read/58825/home/mengenang-prof-pantur-silaban-bapak-ilmu-relativitas-einstein-dari-itb

"Peter Bergmann." Wikipedia. Wikimedia Foundation, November 30, 2022. https://en.wikipedia.org/wiki/Peter_Bergmann

"Proton Decay." Wikipedia. Wikimedia Foundation, December 11, 2022. https://en.wikipedia.org/wiki/Proton_decay

"Relativity v Quantum Mechanics – the Battle for the Universe." The Guardian. Guardian News and Media, November 4, 2015. https://www.theguardian.com/news/2015/nov/04/relativity-quantum-mechanics-universe-physicists

Rehm, Jeremy, and Ben Biggs. "The Four Fundamental Forces of Nature." Space.com. Space, December 23, 2021. https://www.space.com/four-fundamental-forces.html.

Sanchez, N. G. (2019a). The classical-quantum duality of nature including gravity. *International Journal of Modern Physics D*, *28*(3). https://doi.org/10.1142/S021827181950055X

Sanchez, N. G. (2019b). New Quantum Structure of Space-Time. *Gravitation and Cosmology*, *25*(2), 91–102. https://doi.org/10.1134/S0202289319020142

Sanchez, Norma G. "Unifying Quantum Mechanics with Einstein's General Relativity." Research Outreach, November 22, 2022. https://researchoutreach.org/articles/unifying-quantum-mechanics-einstein-general-relativity

"Science Encyclopedia." Science Encyclopedia - JRank Articles, 2022. https://science.jrank.org/.
Simanungkalit, S., Bala, J., Gultom, J. B., & Gultom, J. ". (2004). *Go further with Pantur Silaban*. Fisik.Net. www.fisikanet.lipi.go.id/utama.cgi?artikel&1096193952

Sinusoid, Darya. "The Theory of Relativity: Explained in Simple Terms." Shortform Books, June 15, 2021. https://www.shortform.com/blog/theory-of-relativity-explained/

Siswadi, Anwar. "Fisikawan Pantur Silaban Wafat, Konsisten Meneliti Teori Einstein." Tempo. TEMPO.CO, August 1, 2022. https://tekno.tempo.co/read/1618131/fisikawan-pantur-silaban-wafat-konsisten-meneliti-teori-einstein

Sullivan, Walter. "Physicists to Test Theory on Atoms'slow Decay Test in Lake Erie Mine." The New York Times. The New York Times, December 17, 1979. https://www.nytimes.com/1979/12/17/archives/physicists-to-test-theory-on-atoms-slow-decay-test-in-lake-erie.html

"Theory of Relativity." Wikipedia. Wikimedia Foundation, December 13, 2022. https://en.wikipedia.org/wiki/Theory_of_relativity

Trisetyarso, A., & Silaban, P. (2008). *Degeneracy Breaking of Hydrogen Atom*. http://arxiv.org/abs/0812.4262

"Unified Field Theory." Wikipedia. Wikimedia Foundation, September 15, 2022. https://en.wikipedia.org/wiki/Unified_field_theory

Wells, James. "How Physics at the Roots of Reality Point to a Grand Unified Theory: Aeon Essays." Aeon. Aeon Magazine, May 28, 2021. https://aeon.co/essays/how-physics-at-the-roots-of-reality-point-to-a-grand-unified-theory

Zimmerman Jones, Andrew. "What Was Einstein Talking about with His Theory of Relativity?" ThoughtCo. ThoughtCo, January 8, 2020. https://www.thoughtco.com/einsteins-theory-of-relativity-2699378

Interviews:
Jason Razien, founder and General Manager for Clockworks Technologies

Chapter 9 – Reed

Berger, Michael. "What Are Quantum Dots?" Nanotechnology, July 2, 2021. https://www.nanowerk.com/what_are_quantum_dots.php

Calvet, L. E., Chen, J., Lee, T., & Seabaugh, A. (2022). Mark A. Reed (1955-2021). *Nature Nanotechnology*, *17*(4), 336. https://doi.org/10.1038/s41565-022-01093-5

"Canon : Canon Technology: Canon Science Lab: Optical Computers." Canon Global, 2022. https://global.canon/en/technology/s_labo/light/004/04.html

Neikov, Oleg, and Yefimov, Nikolay A. "Nanopowders." Handbook of Non-Ferrous Metal Powders (Second Edition). Elsevier, January 11, 2019. https://www.sciencedirect.com/science/article/pii/B9780081005439000099.

Fujisawa, T., Hayashi, T., Sasaki, S., Reed, M. A., Randall, J. N., & Luscombex, J. H. (1992). Semiconductor Science and Technology This content was downloaded from IP address 97 Semiconductor quantum dot resonant tunneling spectroscopy. In *Semicond. Sci. Technol* (Vol. 7).

Hahamy, M. "Mark Reed, Director of Undergraduate Studies for Electrical Engineering, Dies at 66." Yale Daily News, May 14, 2021. https://yaledailynews.com/blog/2021/05/07/mark-reed-director-of-undergraduate-studies-for-electrical-engineering-dies-at-66/.

Hughes-Castleberry, Kenna. "Could Quantum Dots Be Made out of Beer by-Products? New Research Says Yes." The Quantum Insider, July 14, 2022. https://thequantuminsider.com/2022/06/29/could-quantum-dots-be-made-out-of-beer-by-products-new-research-says-yes/

"In Memoriam: Mark A. Reed, Professor of Electrical Engineering & Applied Physics." Yale School of Engineering & Applied Science, May 13, 2021. https://seas.yale.edu/news-events/news/memoriam-mark-reed-professor-electrical-engineering-applied-physics

"In Memory of Professor Mark Reed." In memory of professor Mark Reed. Yale University, June 2021. http://yaleseas.com/remembering-mark-reed/

LaPierre, R. R. (2021). In Memoriam - Professor Mark Reed. *Nano Futures*. https://doi.org/10.1088/2399-1984/ac02cf
Mark A. Reed. (1993). 58 QDotsSciAm1993. *Scientific American*, *58*, 118–123.

"Photolithography." Wikipedia. Wikimedia Foundation, November 13, 2022. https://en.wikipedia.org/wiki/Photolithography.\

Rabiee, N., Ahmadi, S., Soufi, G. J., Hekmatnia, A., Khatami, M., Fatahi, Y., Iravani, S., & Varma, R. S. (2022). Quantum dots against SARS-CoV-2: diagnostic and therapeutic potentials. In *Journal of Chemical Technology and Biotechnology* (Vol. 97, Issue 7, pp. 1640–1654). John Wiley and Sons Ltd. https://doi.org/10.1002/jctb.703

Reed, Elizabeth. "Mark Reed Obituary - Monroe, CT." Dignity Memorial, 2021. https://www.dignitymemorial.com/obituaries/monroe-ct/mark-reed-10183846

Reed, M. A., Randall, J. N., & Luscombe, J. H. (1990). Non-equilibrium quantum dots: transport. In *Nanotechnology* (Vol. 1).
Reed, Mark. "Reed Lab at Yale." Reed Lab @ Yale, October 12, 2020. https://www.eng.yale.edu/reedlab/news.html

Woodford, Chris. "Quantum Dots: Introduction to Their Science and Applications." Explain that Stuff, August 30, 2021. https://www.explainthatstuff.com/quantum-dots.html

Interviews:
Dr. Laurie Calver of Centre de Nanosciences et de Nanotechnolgies, at the Université Paris-Saclay in Palaiseau, France
Elizabeth Reed, wife of Mark Reed
Dr. Hong Tang, Professor of Electrical Engineering at Yale University
Mohsen Nami, Yale postdoctoral researcher

Chapter 10 – Jin

"About." W.M. Keck Foundation, August 31, 2022. https://www.wmkeck.org/about/

"BCS Theory." Wikipedia. Wikimedia Foundation, October 27, 2022. https://en.wikipedia.org/wiki/BCS_theory

Bonnyman-Evans, Clay. "Honoring a Visionary, Lost Too Soon." Colorado Arts and Sciences Magazine, March 25, 2019. https://www.colorado.edu/asmagazine/2019/03/25/honoring-visionary-lost-too-soon

"Bose–Einstein Condensate." Wikipedia. Wikimedia Foundation, December 21, 2022. https://en.wikipedia.org/wiki/Bose%E2%80%93Einstein_condensate

"Bose-Einstein Condensate: A New Form of Matter." NIST, November 27, 2017. https://www.nist.gov/news-events/news/2001/10/bose-einstein-condensate-new-form-matter

Calisi, Rebecca. "The Special Challenges of Being Both a Scientist and a Mom." Scientific American Blog Network. Scientific American, March 30, 2018. https://blogs.scientificamerican.com/voices/the-special-challenges-of-being-both-a-scientist-and-a-mom/

Castleberry, Kenna. "Eclipsed Women: The Need for Scientist-Mothers in Popular Culture." The Debrief, November 3, 2021. https://thedebrief.org/eclipsed-women-the-need-for-scientist-mothers-in-popular-culture/

Caughey, Peter. "CU-NIST Scientist Deborah Jin Receives L'Oréal-UNESCO for Women in Science Award." CU Boulder Today, October 22, 2012. https://www.colorado.edu/today/2012/10/22/cu-nist-scientist-deborah-jin-receives-loreal-unesco-women-science-award

Chang, Kenneth. "Lives: Deborah Jin '90 | Princeton Alumni Weekly." Princeton University. The Trustees of Princeton University, 2017. https://paw.princeton.edu/article/lives-deborah-jin-90

"Cindy Regal: Fannie and John Hertz Foundation: Empowering Limitless Progress." Fannie and John Hertz Foundation | Empowering Limitless Progress, December 13, 2022. https://www.hertzfoundation.org/person/cindy-regal/.

Conrad, Kristin. "Deborah Jin Dies at 47." JILA, September 19, 2016. https://jila.colorado.edu/news-events/news/deborah-jin-dies-47-0

Cornell, Eric A, and Carl E. Wieman. "The Bose-Einstein Condensate." *Scientific American*, 1998, 10–15.

Dance, Amber. "Relationships: Sweethearts in science." *Nature* 542, 261–263 (2017). https://doi.org/10.1038/nj7640-261a

"Deborah S. Jin." Wikipedia. Wikimedia Foundation, October 18, 2022. https://en.wikipedia.org/wiki/Deborah_S._Jin.

"Deborah Jin Fellowship." ARC Centre of Excellence for Engineered Quantum Systems, 2021. https://equs.org/deb-jin-fellowships.

DeMarco, B., and D. S. Jin. "Onset of Fermi Degeneracy in a Trapped Atomic Gas." *Science* 285, no. 5434 (1999): 1703–6. https://doi.org/10.1126/science.285.5434.1703

"FAQ – Frequently Asked Questions." NobelPrize.org, August 19, 2022. https://www.nobelprize.org/frequently-asked-questions/#posthumously

"Fermi Gas." Wikipedia. Wikimedia Foundation, November 18, 2021. https://en.wikipedia.org/wiki/Fermi_gas.
"Feshbach Resonance." Wikipedia. Wikimedia Foundation, October 5, 2022. https://en.wikipedia.org/wiki/Feshbach_resonance

The Franklin Institute. "Deborah Jin - 2008 Laureate of the Franklin Institute in Physics." YouTube. YouTube, July 5, 2016. https://www.youtube.com/watch?v=-b88O3c4zVE.

"History of Scientific Women-Deborah Jin." Deborah JIN. History of Scientific Women, 2021. https://scientificwomen.net/women/jin-deborah-48

Institute of Physics. "Newton Lecture 2014: Professor Deborah Jin." YouTube. YouTube, November 11, 2014. https://www.youtube.com/watch?v=y07KNKzB2tY

"JILA through the Years." CU Boulder Today, April 9, 2012. https://www.colorado.edu/today/2012/04/09/jila-through-years

Kino, Lily. "Men's Restrooms Still Outnumber Women's in STEM Buildings." The Independent Florida Alligator, April 15, 2022. https://www.alligator.org/article/2022/04/stem-building-bathrooms-show-gender-inequality

Kuta, Sarah. "Deborah Jin Remembered by Boulder Colleagues as 'One of the Great Atomic Physicists of Our Day'." Boulder Daily Camera. Boulder Daily Camera, September 19, 2016. https://www.dailycamera.com/2016/09/19/deborah-jin-remembered-by-boulder-colleagues-as-one-of-the-great-atomic-physicists-of-our-day/
.

L'Oreal USA For Women in Science. "Deborah S Jin, 2013 L Oréal-UNESCO Awards Laureate for North America." YouTube. YouTube, April 26, 2013. https://www.youtube.com/watch?v=wYKjjoUIEN0

Ni, Kang-Kuen, Markus Greiner, and Silke Ospelkaus. "Deborah S. Jin 1968-2016: Trailblazer of Ultracold Science." Proceedings of the National Academy of Sciences of the United States of America. U.S. National Library of Medicine, January 31, 2017. https://www.ncbi.nlm.nih.gov/pmc/articles/PMC5293022/

"The Nobel Prize in Physics 2001." NobelPrize.org, 2022. https://www.nobelprize.org/prizes/physics/2001/popular-information/

Padavic-Callaghan, Karmela. "Deborah Jin Engineered New Quantum States of Matter - Twice." Massive Science, August 7, 2020. https://massivesci.com/articles/deborah-jin-science-hero-ultracold-quantum-fermions/

Phillips, Julie. *JILA: The First 50 Years*. Boulder, CO: JILA, 2012.

"Physics Degrees Earned by Women." American Physical Society. APS, IPEDS, and NSF, 2020. https://www.aps.org/programs/education/statistics/womenphysics.cfm

Powell, Kendall. "The Parenting Penalties Faced by Scientist Mothers." Nature News. Nature Publishing Group, July 20, 2021. https://www.nature.com/articles/d41586-021-01993-x

"Prominent Nist and Jila Physicist Deborah Jin Passes Away." NIST, January 8, 2018. https://www.nist.gov/news-events/news/2016/09/prominent-nist-and-jila-physicist-deborah-jin-passes-away

Randeria, Mohit, and Edward Taylor. "BCS-Bec Crossover and the Unitary Fermi Gas." arXiv.org, April 8, 2014. https://arxiv.org/abs/1306.5785

Regal, Cindy., Greiner, Markus., Giorgini Stefano., Holland Murray., and Jin, Deborah. "Momentum Distribution of a Fermi Gas of Atoms in the BCS-Bec Crossover." *Physical Review Letters* 95, no. 25 (2005). https://doi.org/10.1103/physrevlett.95.250404

Regal, Cindy., & Ye, Jun. (2016). "Deborah S. Jin (1968-2016)." *Science* (Vol. 354, Issue 6313, p. 709). American Association for

the Advancement of Science.
https://doi.org/10.1126/science.aal2545

Toon, John. "Bose-Einstein Condensates Evaluated for Communicating among Quantum Computers." Bose-Einstein Condensates Evaluated for Communicating Among Quantum Computers | School of Physics. Georgia Tech University, 2013. https://physics.gatech.edu/news/bose-einstein-condensates-evaluated-communicating-among-quantum-computers

Villalba, Constanza. "Partners in Science, Part One: Negotiating the Academic Job Market as Postdocs." Science, 1999. https://www.science.org/content/article/partners-science-part-one-negotiating-academic-job-market-postdocs

Weil, Martin. "Deborah Jin, Government Physicist Who Won MacArthur 'Genius' Grant, Dies at 47." The Washington Post. WP Company, September 20, 2016. https://www.washingtonpost.com/national/health-science/deborah-jin-government-physicist-who-won-macarthur-genius-grant-dies-at-47/2016/09/20/c2deb8ce-7ed9-11e6-8d13-d7c704ef9fd9_story.html

Interviews:
John Bohn, JILA Fellow
Eric Cornell, JILA and NIST Fellow
Jun Ye, JILA and NIST Fellow
Cindy Regal, JILA Fellow
Brian DeMarco, Professor at the University of Illinois
Carrie Weidner, Quantum Engineering Lecturer at University of Bristol

Conclusion:

Ferrie, Chris. *Quantum Bullshit: How to Ruin Your Life with Advice from Quantum Physics*. Naperville, IL: Sourcebooks, 2023.

Interviews:

Dr. Mariia Mykhailova, Principal Software Engineer at Microsoft Quantum

INDEX

Made in United States
Orlando, FL
04 November 2026

90204189R00146